Essential Concepts in Toxicogenomics

METHODS IN MOLECULAR BIOLOGY™

John M. Walker, SERIES EDITOR

METHODS IN MOLECULAR BIOLOGY™

Essential Concepts in Toxicogenomics

Edited by

Donna L. Mendrick

Gene Logic Inc,
Gaithersburg, MD, USA

and

William B. Mattes

The Critical Path Institute, Rockville, MD, USA

 Humana Press

Editors
Donna L. Mendrick
Gene Logic Inc,
Gaithersburg, MD, USA
dmendrick@genelogic.com

William B. Mattes
The Critical Path Institute,
Rockville, MD, USA
wmattes@C-Path.org

Series Editor
John M. Walker
Professor Emeritus
School of Life Sciences
University of Hertfordshire
Hatfield
Hertfordshire AL10 9AB, UK
john.walker543@ntlworld.com

ISBN: 978-1-58829-638-2   e-ISBN: 978-1-60327-048-9

Library of Congress Control Number: 2008921701

Cover illustration: Chapter 2, Fig. 1.

Printed on acid-free paper

9 8 7 6 5 4 3 2 1

springer.com

Preface

The field of toxicogenomics is moving rapidly, so it is impossible at the time of this writing to compile a classic methods textbook. Instead, we chose to identify experts in all aspects of this field and challenged them to write reviews, opinion pieces, and case studies. This book covers the main areas important to the study and use of toxicogenomics. Chapter 1 speaks to the convergence of classic approaches alongside toxicogenomics. Chapter 2 deals with the usefulness of toxicogenomics to identify the mechanism of toxicity. Chapter 3 calls attention to the issues that affect the quality of toxicogenomics experiments, as well as the implications of using microarrays as diagnostic devices. The need for appropriate statistical approaches to genomic data is discussed in Chapter 4, and Chapters 5 and 6 describe the use of genomic data to build toxicogenomic models and provide insights from the approaches of two companies. The important topic of storing the data generated in such experiments and the correct annotation that must accompany such data is considered in Chapter 7. The discussion in Chapter 8 speaks to the use of toxicogenomics to identify species similarities and differences. Chapters 9 and 10 deal with the use of genomics to identify biomarkers within the preclinical and clinical arenas. Biomarkers will only be useful if the community at large accepts them as meaningful. Consortia are important to drive this function, and Chapter 11 discusses current efforts in this area. Last but not least, Chapter 12 presents a perspective on the regulatory implications of toxicogenomic data and some of the hurdles that can be seen in its implication in GLP studies. Although this book tends to focus on pharmaceuticals, the issues facing toxicology are shared by the chemical manufacturers, the tobacco industry, and their regulators. We want to thank our contributors for their generous time and energy in providing their insights. Sadly, we must note the unexpected passing of one of our authors, Dr. Joseph Hackett of the FDA. Joe's contribution serves as a testimony to his accomplishments in this field, and his insight will be missed in the years to come.

Donna L. Mendrick

William B. Mattes

Contents

Contributors

SCOTT A. BARROS • *Toxicology, Archemix Corp., Cambridge, Massachusetts*

ERIC A. BLOMME • *Department of Cellular and Molecular Toxicology, Abbott Laboratories, Abbott Park, Illinois*

WAYNE R. BUCK • *Department of Cellular and Molecular Toxicology, Abbott Laboratories, Abbott Park, Illinois*

LYLE D. BURGOON • *Department of Biochemistry & Molecular Biology, Michigan State University, East Lansing, Michigan*

JOHN N. CALLEY • *Department of Integrative Biology, Eli Lilly and Company, Greenfield, Indiana*

KELLYE K. DANIELS • *Department of Toxicogenomics, Gene Logic Inc., Gaithersburg, Maryland*

MARC F. DeCRISTOFARO • *Biomarker Development, Novartis Pharmaceuticals Corporation, East Hanover, New Jersey*

JOSEPH J. DeGEORGE • *Laboratory Sciences and Investigative Toxicology, Merck & Co Inc, West Point, Pennsylvania*

MICHAEL ELASHOFF • *Department of BioStatistics, CardioDx, Palo Alto, California*

JOSEPH HACKETT • *Office of Device Evaluation, Center for Devices and Radiological Health, U.S. Food and Drug Administration, Rockville, Maryland*

RORY B. MARTIN • *Drug Safety and Disposition, Millennium Pharmaceuticals, Cambridge, Massachusetts*

WILLIAM B. MATTES • *Department of Toxicology, The Critical Path Institute, Rockville, Maryland*

DONNA L. MENDRICK • *Department of Toxicogenomics, Gene Logic Inc., Gaithersburg, Maryland*

MARK W. PORTER • *Department of Toxicogenomics, Gene Logic Inc., Gaithersburg, Maryland*

TIMOTHY P. RYAN • *Department of Integrative Biology, Eli Lilly and Company, Greenfield, Indiana*

FRANK D. SISTARE • *Laboratory Sciences and Investigative Toxicology, Merck & Co Inc., West Point, Pennsylvania*

KAROL L. THOMPSON • *Division of Applied Pharmacology Research, Center for Drug Evaluation and Research, U.S. Food and Drug Administration, Silver Spring, Maryland*

JEFFREY F. WARING • *Department of Cellular and Molecular Toxicology, Abbott Laboratories, Abbott Park, Illinois*

TIMOTHY R. ZACHAREWSKI • *Department of Biochemistry and Molecular Biology, Michigan State University, East Lansing, Michigan*

Color Plates

Color plates follow p. 112.

1

Toxicogenomics and Classic Toxicology: How to Improve Prediction and Mechanistic Understanding of Human Toxicity

Donna L. Mendrick

Summary

The field of toxicogenomics has been advancing during the past decade or so since its origin. Most pharmaceutical companies are using it in one or more ways to improve their productivity and supplement their classic toxicology studies. Acceptance of toxicogenomics will continue to grow as regulatory concerns are addressed, proof of concept studies are disseminated more fully, and internal case studies show value for the use of this new technology in concert with classic testing.

Key Words: hepatocytes; hepatotoxicity; idiosyncratic; phenotypic anchoring; toxicogenomics; toxicology.

1. Introduction

The challenges facing the field of toxicology are growing as companies demand more productivity from their drug pipelines. The intent of this chapter is to identify the issues facing classic approaches to nonclinical toxicity testing, the cause of the deficiencies, and ways in which toxicogenomics can improve current *in vitro* and *in vivo* testing paradigms. The public at large continues to exert pressure on the pharmaceutical industry to develop new drugs yet are intolerant of safety issues and the high cost of drugs when they reach the market. This is setting up a "perfect storm" with a recognized decrease in productivity in this industry, continual increase in costs of developing new drugs, and rising attrition rates due to nonclinical and clinical safety failures (*1–4*).

From: *Methods in Molecular Biology, vol. 460: Essential Concepts in Toxicogenomics*
Edited by: D. L. Mendrick and W. B. Mattes © Humana Press, Totowa, NJ

2. Classic Toxicology

Those in the field of drug and chemical development know of the multitude of compounds to which humans were never exposed either in the clinic or in the environment because of obvious toxicity seen in preclinical species. However, it is well-known that classic testing in animals is not infallible. A study done by a group within the Institutional Life Sciences Institute (ILSI) illustrates the problem *(5)*. Twelve companies contributed data on 150 compounds that have shown toxicity in humans of a significant enough nature to warrant one of four actions: (1) termination, (2) limitation of dosage, (3) need to monitor drug level, or (4) restriction of target population. The group compared the human toxicities with the results of the classic toxicity employed for each drug. They found that only ~70% of these toxicities could be predicted in classic animal testing even when multiple species, primarily the rat (rodent) and dog (nonrodent), were employed. The dog was better than the rat in predicting human toxicity (63% vs. 43%, respectively), with the success rate varying depending on the human target organ. However, escalating concerns regarding the use of animals in medical research, the amounts of compound required for such large animals, and the cost of such studies prevents this species from being used as the first species or in sufficient numbers to detect subtle toxicities. The exact failure rate due to toxicity and the time of its detection continues to be the subject of study because only by understanding the problem can one begin to propose solutions. Authors tend to report somewhat different findings. The drugs terminated because of human toxicities evaluated in the ILSI study *(5)* failed most often during Phase II (**Fig. 1**). Suter et al. at Roche

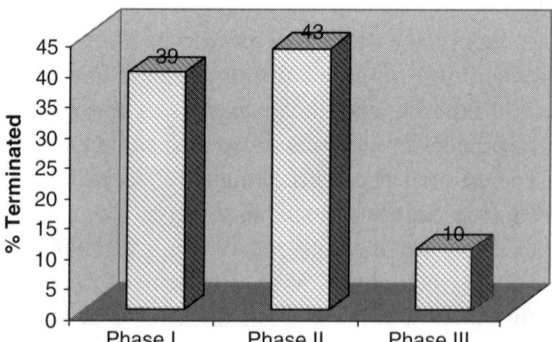

Fig. 1. Data illustrating the termination rate of compounds due to human toxicity during clinical trials. (Data adapted from Olson, H., Betton, G., Robinson, D., Thomas, K., Monro, A., Kolaja, G., et al. 2000. Concordance of the toxicity of pharmaceuticals in humans and in animals. *Regul. Toxicol. Pharmacol.* **32**, 56–67.)

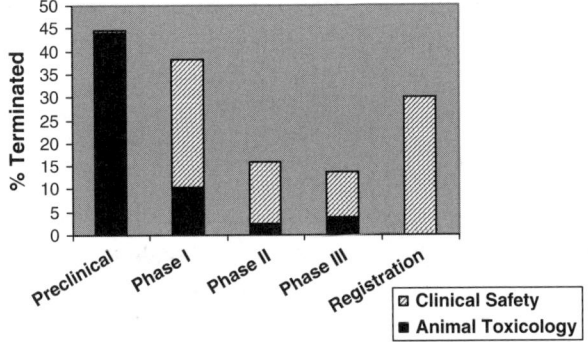

Fig. 2. Failure rate due to animal toxicity and human safety during the development pipeline. (Data adapted from Suter, L., Babiss, L.E., and Wheeldon, E.B. 2004. Toxicogenomics in predictive toxicology in drug development. *Chem. Biol.* **11**, 161–171.)

(6) examined the failure rate of compounds from preclinical to registration and divided safety failures into animal toxicity and human toxicities. Their work found the highest failure rates due to human safety in Phase I and registration (**Fig. 2**). Note that both studies found a high rate of failure in Phase II or beyond, a costly scenario. Dimasi and colleagues have examined financial models of drug development and have estimated the savings of terminating unsafe compounds earlier within the clinical trial paradigm *(7)*. For example, if 25% of the drugs that will fail in Phase II were discontinued in Phase I, the clinical cost savings alone per approved drug would be $13 million to $38 million dollars. Obviously, the cost savings will be greater if one could prevent such a drug from even entering clinical trials by improving preclinical detection and/or by failing it earlier within the clinical testing phase (e.g., Phase I vs. Phase III). The Food and Drug Administration's (FDA) Critical Path Initiative (www.fda.gov/oc/initiatives/criticalpath/) quotes one company executive as saying clinical failures due to hepatotoxicity had cost the company more than $200 million per year over the past decade. Clearly, there are many financial incentives to address the issue of safety.

3. Toxicogenomics

Many in the field have written excellent opinion pieces and reviews on the use of toxicogenomics in drug discovery and development and in the chemical/agrochemical sectors. Toxicogenomics is used in three areas: predictive applications for compound prioritization, mechanistic analyses for compounds with observed toxicity, and biomarker identification for future screens or to develop biomarkers useful in preclinical and/or clinical studies. Though impractical to list all of the relevant publications, a few excellent articles on toxicogenomics are provided *(2,3,6,8–15)*.

3.1. Study Designs

As with all scientific endeavors, to answer the questions being posed it is important to have an optimal study design. Genomics tends to be somewhat expensive, so understandably some try to downsize the experimental setting. Unfortunately, that may prevent hypothesis generation or evaluation of a preexisting theory. As an example, if one is trying to form a hypothesis as to the mechanism of injury induced by a compound, sampling tissue only at the time of such damage may prevent evaluation of the underlying events that started the pathologic processes. Similarly, sampling only one time point will inhibit the fullest evaluation of the dynamic processes of injury and repair. In classic toxicology testing, one would not claim with certainty that a compound is not hepatotoxic if one saw no elevation of serum alanine aminotransferase (ALT) or histologic change in the liver in a snapshot incorporating only one time point and dose level. Likewise, one does not pool blood from all animals and perform clinical pathology on such. Unfortunately, some have approached toxicogenomics in this manner (using restrictive study designs and pooled RNA samples) and then felt betrayed by the lack of information. This does not mean to suggest that all toxicogenomic studies must be all-encompassing as long as the investigator understands beforehand the limits of his or her chosen design. One approach might be to collect samples from multiple time points and doses and triage the gene expression profiling to determine the most important study groups within that experimental setting. Establishment of the appropriate dose is important as well. Classic toxicology endeavors use dose escalation until one sees a phenotypic adverse event such as changes in classic clinical pathology, histology, body weight, and so forth. An anchoring of the dose used for toxicogenomic studies also must be employed and contextual effects of such doses understood. Doses that severely affect body weight likely induce great stress upon the animal, and this must be taken into account if this phenotypic anchor is followed. A recent paper by Shioda et al. studied effects of xenoestrogens in cell culture and explored the relationship of doses to transcriptional profiles *(16)*. This work suggests that doses chosen for equivalent cellular responses highlight the differences between compounds while those selected based on the compounds' action on a particular gene reveal mostly similarities between the compounds. Additional work remains to be done to determine if this conclusion can be extrapolated to other compound types and *in vivo* environments, but, at a minimum, this report reinforces the need to understand the chosen study design and fully explain criteria used for dose selection.

3.2. Genomic Approaches Can Clarify Basic Husbandry Issues

Genomics enables detection of toxicity parameters as well as differences in animal husbandry. In many cases, the study design may call for food restriction or animals may be accidentally deprived of food. Genomic analysis can detect such events as shown in **Fig. 3** and **Fig. 4**. Studies in Gene Logic's (Gaithersburg, MD) ToxExpress® database were used for the analysis. In **Fig. 3**, the data from the probe sets (~8800) on the Affymetrix Rat Genome U34A GeneChip® microarray (Santa Clara, CA) were subjected to a principal components analysis (PCA). Such a test illustrates underlying differences in the data in a multidimensional picture. For ease in viewing, two-dimensional graphical representations are provided. In **Fig. 3**, the data from all probe sets were used, and, even with the accompanying noise when so many parameters are measured, one can see differentiation of the groups particularly if one combines the x and y axes, accounting for 39% of the gene expression variability.

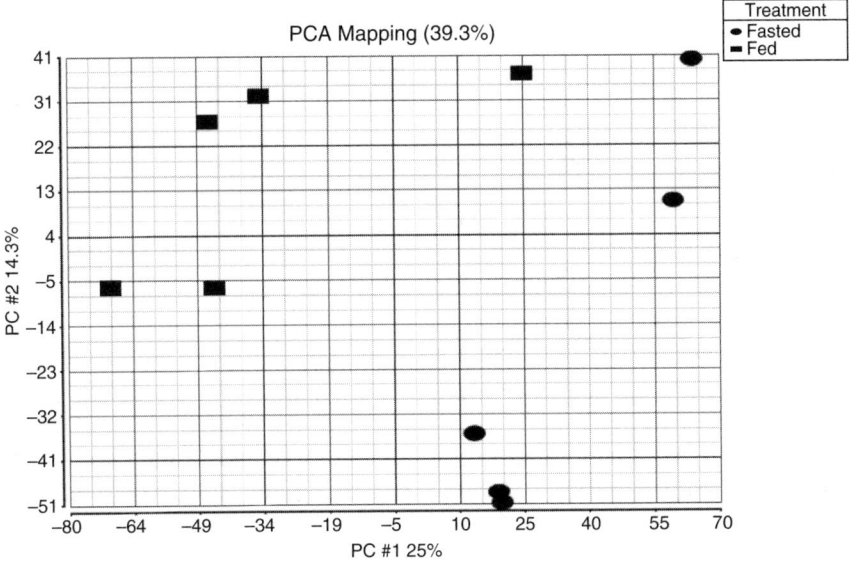

Fig. 3. A PCA using all genes on the Rat Genome U34A microarray illustrates the differentiation on a genomic basis between rats fasted for 24 h versus those rats that had food *ad libitum*. Use of all genes on the array is accompanied by noise and yet one obtains reasonable separation using the x axis and better discrimination if one employs both x and y axes. Such findings illustrate the ability of gene expression to provide insight into animal husbandry.

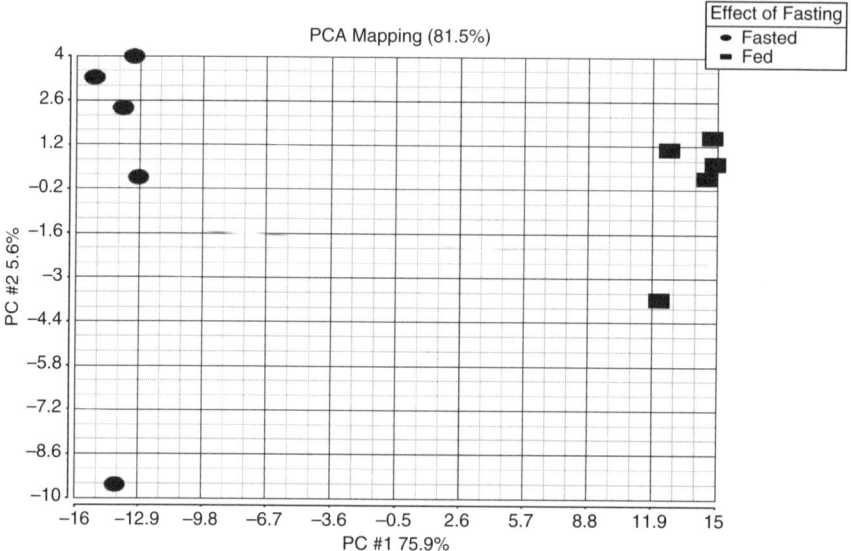

Fig. 4. An analysis filter was applied to identify differentially regulated genes. The filter has a cutoff as follows: fold-change ≥ 1.8 with t-test p-value $<.05$ and $\geq 90\%$ present in reference or experimental group with mean avg. diff. >40. This resulted in 281 genes identified as dysregulated between fed and fasted rats. Almost all of the variation is captured in PC #1 (75.9%) and the groups are more clearly separated than seen when all genes were used as shown in **Fig. 3**.

When genes that were differentially regulated among the fed and fasted animals were chosen, the gene list was reduced from >5000 to 281. The results in **Fig. 4** demonstrate a complete discrimination of these rats with the x axis accounting for 76% of the variability.

Genomics can be used to discriminate strain and gender as well. In the former case, female rats of Sprague-Dawley (SD) or Wistar origin were compared. Although evaluation using all genes discriminated these strains (data not shown), selection of differentially regulated genes resulted in a clearer separation as shown in **Fig. 5**. Because both strains are albino, one could envision using a genomics approach should there be a potential mix-up of strains in the animal room.

What is likely less surprising is the ability to categorize gender based on gene expression findings. Although it is usually easy to identify the gender of rats by physical examination alone, one could envision the use of a genomic approach to study the feminization of male rats under drug treatment or vice versa. As shown in **Fig. 6**, a PCA employing all genes discriminates between genders although the first two axes capture only ~23% of the variation suggesting there

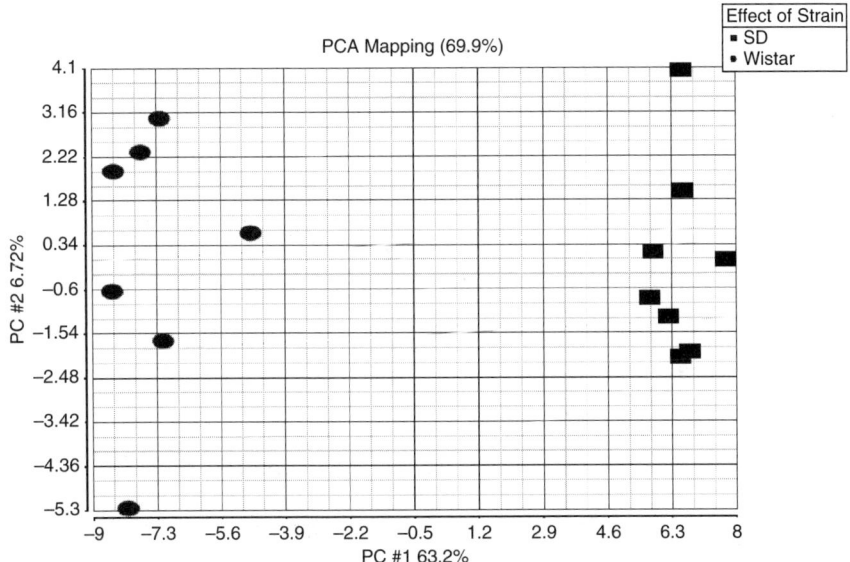

Fig. 5. A PCA was performed with 83 genes that were differentially expressed between normal female Sprague-Dawley (SD) or Wistar rats. Such strains are clearly discriminated using a genomic approach.

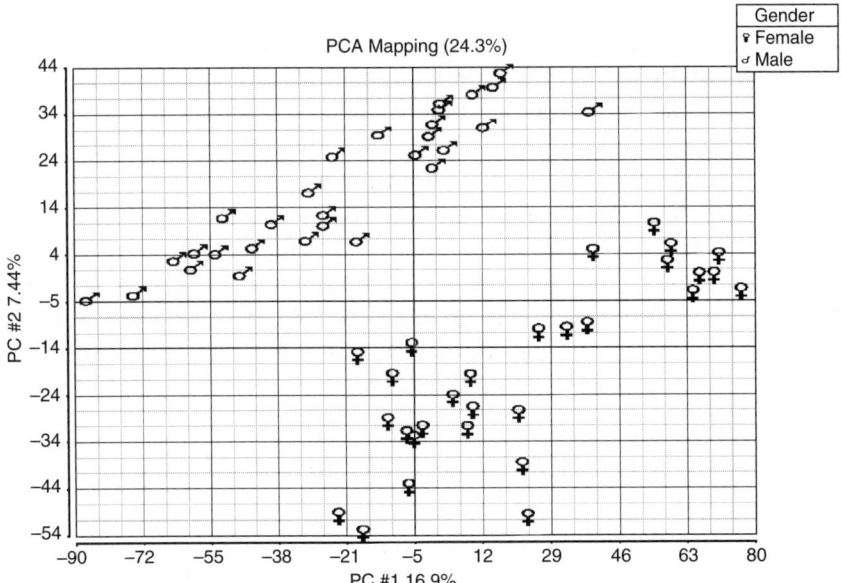

Fig. 6. A PCA was performed with all genes from male and female rat livers. There is an overall separation particularly if one combines the variability shown along the x and y axes.

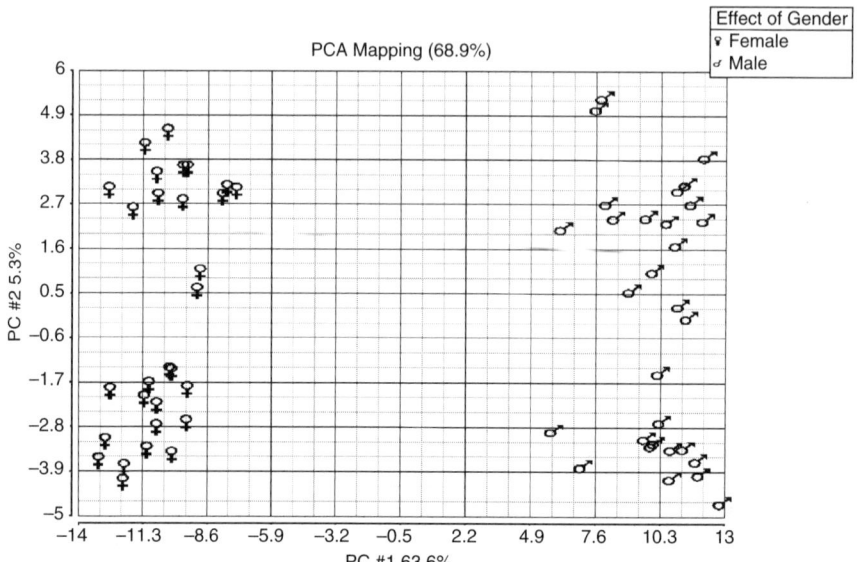

Fig. 7. A PCA was performed with 175 genes that were differentially expressed between the genders. In this case, the majority of difference was captured in the first axis resulting in a very clear separation of the genders.

are many subtle effects, a point not likely to be argued by many humans. Using genes differentially regulated, however, found that 175 genes can account for more than 63% of variation in the first axis alone as shown in **Fig. 7**.

In the three cases described above (food deprivation, strain, and gender), differentially regulated genes were identified that enabled clear categorization between the groups. However, the reverse is also true. Removal of genes that identify such differences from the analysis can enable other differences to become more apparent. Such has been accomplished with the predictive models built as part of our ToxExpress program specifically to avoid such confounding variables and enable such models to work well when female or male rats are used of various strains and independent of fasting.

3.3. Toxicogenomics Can Augment Understanding of Classic Toxicology Findings

Toxicogenomics can add value to classic testing when one animal appears to have an aberrant finding. It is recognized that preclinical testing of drugs is a complicated process and mistakes can happen. If a rat did not exhibit any signs of toxicology unlike its cohorts, it would be informative to differentiate dosing errors from true differences in toxicity responses. One could monitor the blood to determine if the drug was detectable, but that would depend on its clearance. Alternatively, one could explore gene expression as a subtle detection method. **Figure 8** illustrates how similarity in gene expression can identify the treatment given. If nine rats received carbon tetrachloride and the tenth rat was left untreated, the gene expression of the latter would appear similar to that of untreated rats as shown in the upper left corner in this mock illustration.

As seen with these examples, molecular approaches can be more sensitive in terms of discriminating animal husbandry, strain differences, and so forth, than parameters monitored in classic toxicity testing. Microarrays monitor tens of thousands of events (expression levels of genes and Expressed Sequence Tags (ESTs)), whereas classic toxicity testing has been estimated to measure approximately 100 parameters *(17)*. From a strictly statistical approach, one can envision that more knowledge would lead to improved decision making although the challenge of removing the noise when monitoring so many events is real. Arrays provide information on the changes in individual genes, and from this one can then hypothesize on the effects on biological pathways, and

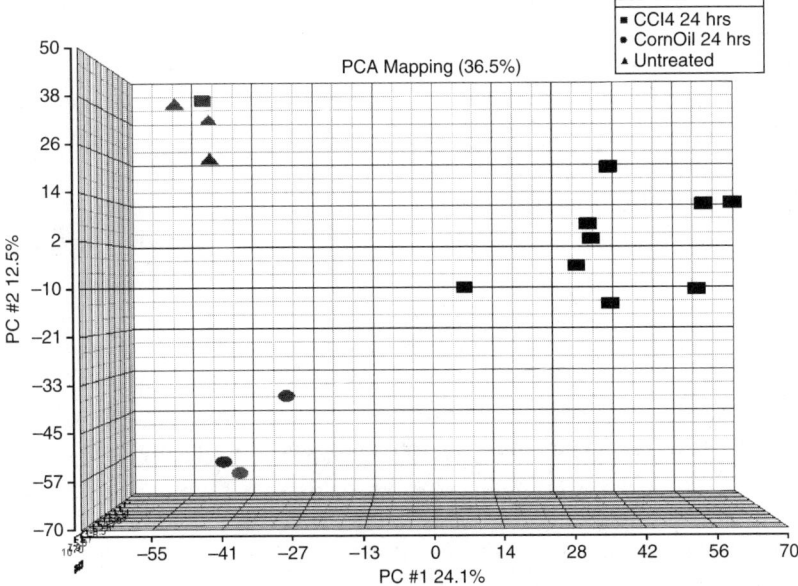

Fig. 8. A PCA illustrates the greatest source of difference at the gene expression level between the rats. The first component (PC #1) accounts for 24% of the variability, and rats treated with carbon tetrachloride are clearly delineated from those receiving vehicle or left untreated with the exception of the one rat in the upper left. The second component, PC #2, accounts for 12.5% of the variability and discriminates even untreated from vehicle-treated rats further illustrating the sensitivity of this approach.

so forth, a level of understanding not provided in classic approaches. One could use the analogy of the sensitivity of electron versus light microscopy. Disease processes such as minimal change nephropathy induce severe functional alterations (proteinuria) in the human kidney yet are extremely difficult to visualize with light microscopy. However, the morphologic changes associated with this proteinuria are clearly seen in such diseased kidneys if one uses electron microscopy, a more sensitive technique. At the ultrastructural level, a flattening of visceral epithelial cells upon the glomerular basement membrane is clearly visible, and such a process is known to be associated with proteinuria in preclinical species and in humans.

Several recent papers highlight the improved sensitivity of a genomics approach with classic examples of compounds that produce adverse events in rats. Heinloth and her colleagues at the National Institute of Environmental Health and Safety (NIEHS) have examined the dose response relationship of rat liver to acetaminophen *(18)*. They studied multiple doses in the rat using two (they called these "subtoxic") doses that failed to cause changes monitored with classic approaches. However, when ultrastructural methods were employed, some toxicity-associated mitochondrial changes were observed at one of these "subtoxic" doses. Evaluation of gene expression changes found, as a consequence of acetaminophen toxicity, a loss in cellular energy even at such "subtoxic" doses. As the dose was increased, so was the magnitude of changes observed, and this was accompanied by alterations in associated genes. They concluded that gene expression profiling may provide more sensitivity for detecting adverse effects in the absence of the occurrence of overt toxicity.

Another recent paper explored a time savings approach. Nie and his colleagues at Johnson and Johnson Pharmaceutical Research and Development examined changes in gene expression in the rat liver at 24 h after exposure to nongenotoxic carcinogens and control compounds *(19)*. They identified six genes that predicted 89% of nongenotoxic carcinogens after 1 day of exposure instead of the 2 years of chronic dosing normally required for such cellular changes to be clear. They also mined the gene expression results to find biologically relevant genes for a more mechanistic approach to understanding nongenotoxic carcinogenicity. Together the work by Heinloth et al. and Nie et al. highlight the sensitivity of a genomics approach to detect compounds that will elicit classic adverse events in rats either at higher doses or exposure times.

Toxicogenomics can be employed to identify species-specific changes providing support for safety claims in humans. Peroxisome proliferator activated receptor alpha (PPAR-α) agonists have been widely studied as they have been found to be clinically useful as hypolipidemic drugs yet induce hepatic tumors in rats. It is now appreciated that such an effect has little safety risk to humans and that genes can be identified in rat liver and hepatocytes that enable discrimination of this mechanism. As suggested by Peter Lord and his colleagues at Johnson and Johnson, this could be incorporated into the safety evaluation of novel compounds by employing expression profiling of rat liver or hepatocytes exposed to the new drug. Similarities between this novel compound and known PPAR-α agonists could provide a safety claim for the former's species-specific effects *(15)*.

In some cases, differences in species responses to drugs may rely on dissimilarities in drug metabolism enzymes. A study was performed using data in Gene Logic's ToxExpress database. Expression levels of genes involved in glutathione metabolism were compared among normal tissues and genders of the rat, mouse, and canine. As expected, expression levels varied among genes and tissues. Little differences were seen among male and female animals but large effects were seen among species. If differential gene expression does translate to enzymatic activities differences between species, it would be expected to impact drug responses. Such information could provide guidance into species selection for toxicity testing *(20)*. Chapters later in this book discuss species differences in more detail.

3.4. Use of Toxicogenomics to Detect Idiosyncratic Compounds

A well-accepted definition of the term *idiosyncratic*, particularly on a compound by compound basis, is hard to find as individuals use different criteria. Some depend on incidence alone, some on the inflammatory response induced, some on a lack of dose response relationship, and so forth. However, universal to most users' definitions is the lack of classic changes observed in preclinical species. To complete the analogy started above in discussing the sensitivity differences between light and electron microscopy, one could suppose that idiosyncratic compounds elicit some effects at the gene and protein level in rats but these events do not lead to changes severe enough to induce classic phenotypic changes in this commonly used testing species. The failure to cause overt injury in the rat might be due to (1) the inability of this species to metabolize the drug into the toxic metabolite responsible for human injury, (2) quantitative differences whereby rats generate the toxic metabolite but at lower levels than seen in humans, (3) a superior ability of the rat to detoxify toxins, and (4) the failure of rats to respond in an immune-mediated manner to adduct formation of intrinsic cell surface proteins. Many if not all compounds today are commonly screened for metabolism differences between species and their development discontinued if such metabolic specific responses are seen. Even so, the postmarketing occurrence of drug hepatotoxicity is a leading cause of regulatory actions and further erodes the pharmaceutical pipeline. The field of drug development needs new methods to identify such drugs earlier in discovery and development that can inform compound selection, position drugs to areas of unmet clinical need, and influence preclinical and clinical trial designs in a search for earlier biomarkers of liver injury induced by such a compound.

Several groups including those from Millennium (**Chapter 5** and Ref. *21*), AstraZeneca *(22)*, and Gene Logic (**Chapter 6**) have reported success in (a) training predictive models using all genes and ESTs being monitored on the array platform and (b) classifying the compounds based on their potential to induce toxicity in humans including idiosyncratic compounds. The result is sets of genes and ESTs that may not be easily translated into mechanistic understanding on their own yet show robust predictive ability. Such an approach has enabled the prediction of idiosyncratic hepatotoxicants from rats and rat hepatocytes exposed to such agents. **Figure 9** illustrates the predictive ability of models built at Gene Logic to identify idiosyncratic compounds as potential human hepatotoxicants using gene expression data obtained from *in vivo* and *in vitro* exposure to prototypical compounds. The classification of individual compounds as idiosyncratic tends to be controversial, so to remove any internal bias, the statistics were compiled using marketed compounds classified as idiosyncratic by Kaplowitz *(23)*. Kaplowitz subclassifies idiosyncratic compounds as allergic or nonallergic. As can be seen in **Fig. 9**, toxicogenomic predictions performed from *in vivo* rat exposure are equally accurate regardless of whether or not the idiosyncratic compound induces an allergic responses in the human liver, using Kaplowitz's classification. In contrast, the model built from rat hepatocyte data does somewhat better with compounds that

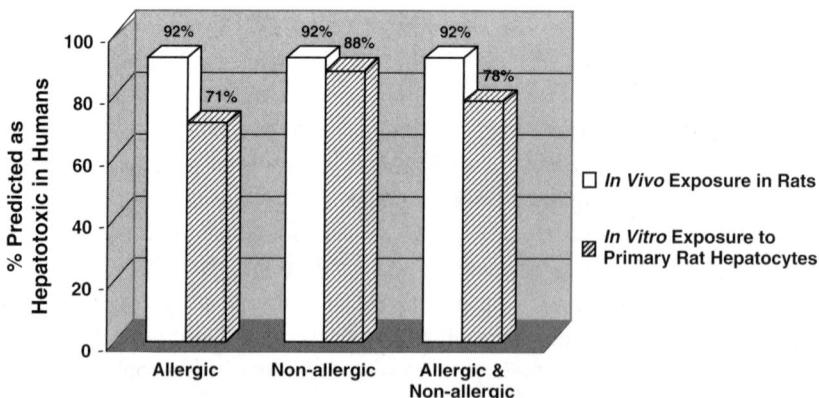

Fig. 9. Compounds described as idiosyncratic by Kaplowitz *(23)* were studied. He discriminates between drugs that induce allergic-type reactions from those that do not, and the results of the predictive toxicogenomic modeling are shown here.

cause nonallergic responses in humans. It should be remembered that idiosyncratic compounds do not elicit classic phenotypic changes in rats and other nonclinical species, yet the majority are detected using a predictive toxicogenomics approach in rat liver and primary rat hepatocytes. How is it possible for genomics to predict something in the rat that is not seen at a classic phenotypic level?

3.4.1. Case Studies of Idiosyncratic Hepatotoxicants

For most of the idiosyncratic drugs on the market today, little is known about their ability to induce hepatotoxicity in humans while avoiding classic detection in preclinical species. Felbamate is an exception, and it has been researched extensively. It has been reported that its metabolism is quantitatively different between species and this is believed to be responsible for the failure of classic testing to detect its potential to induce human hepatotoxicity. Researchers have reported that felbamate is metabolized to a reactive aldehyde (atropaldehyde) and that rats produce far less of this toxin than do humans *(24)*. However, because rats do produce some level of this toxic metabolite, it raises the possibility that a technology more sensitive than classic testing (e.g., histopathology) may detect subtle signs of toxicity in rat. To test this hypothesis, an *in vivo* study done at Gene Logic employed 45 rats in groups of five per dose and time point. Animals were dosed daily by oral gavage with vehicle, low-dose or high-dose felbamate. Groups were sacrificed at 6 h and 24 h after one treatment and 14 days after daily dosing. Histology of the livers was normal. One of the high-dose–treated rats sacrificed after 14 days of exposure exhibited an ALT level above the normal reference range, and the entire group had a minor but statistically significant elevation above the control group. The overall conclusion of the board-certified veterinary pathologist was that a test article effect was unclear. Because this drug failed to demonstrate hepatotoxicity in the classic testing employed for registration, this result was not surprising. However, it does raise the issue of whether individual rats are responding in a classic sense to idiosyncratic compounds, and these results are not being interpreted correctly as suggested by retrospective analysis of some idiosyncratic compounds by Alden and colleagues at Millennium Pharmaceuticals *(25)*. To explore the potential effects of felbamate on gene expression, livers were collected from all rats in this study, processed using Affymetrix GeneChip microarrays, and the data passed through the predictive toxicogenomics models built at Gene Logic. The results of the models, illustrated in **Fig. 10**, were correct in identifying felbamate as a potential human hepatotoxicant, a potential hepatitis-inducing agent in humans *(26,27)*, and an agent that will cause liver enlargement in rats

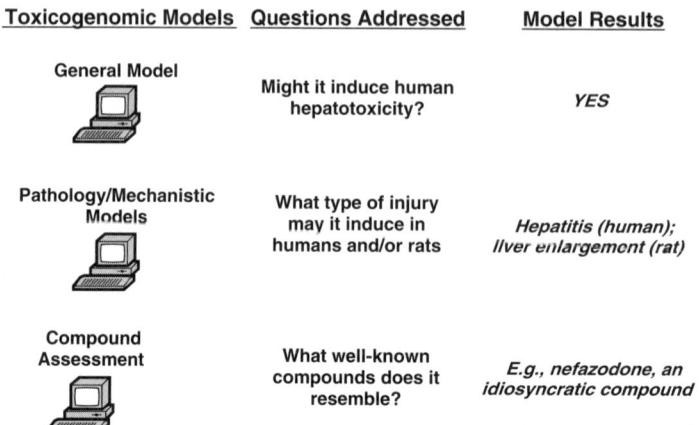

Fig. 10. Predictive toxicogenomic models were built at Gene Logic and validated internally and with customer-provided blinded data. The results of the genomic data obtained from felbamate-treated rat liver are shown here.

(28). The gene expression data from such treated rats showed some similarity to other idiosyncratic drugs such as nefazodone.

Another interesting idiosyncratic drug is nefazodone. This oral antidepressant was approved in 1994 in the United States, but postmarket cases of idiosyncratic adverse reactions were reported that resulted in 20 deaths. A black box warning was added to the label in 2001 and the drug withdrawn in 2004 *(23,29–34)*. The only hepatic effect reported in rats was a dose-related (50, 150, 300 mg kg^{-1} day^{-1}) increase in liver weight reported in a 13-week rat study (SBA NDA 20-152). Several mechanisms for its toxicity have been proposed. The first hypothesis regarding its toxicity is a generation of reactive metabolites via bioactivation, which are capable of covalently modifying CYP3A4, a drug-metabolizing enzyme it is metabolized by and inhibits *(35–37)*. The second theory is that nefazodone compromises biliary elimination, resulting in an increase in drug and bile acids retained in the liver over time *(34,38)*. These authors have found a transient increase in serum bile acids in rats suggesting that rats are responding to this drug inappropriately but can recover from its toxicity prior to developing classic phenotypic changes.

To investigate whether gene expression changes might be able to detect nefazodone as a potential human hepatotoxicant in the absence of classic changes in the rat, a study was performed using doses of 0, 50, and 500 mg kg^{-1} day^{-1} given by oral gavage. Note that the high dose is only 67% higher than what the manufacturer used in its 13-week rat study in which liver enlargement was reported. Groups of rats were sacrificed 6 h and 24 h after the first dose and 7 days after daily dosing. Clinical pathology and liver histopathology revealed no clear evidence of liver toxicity as was expected of this idiosyncratic compound. The results from toxicogenomic modeling were similar to that of felbamate, namely that it is a potential human hepatotoxicant that can cause hepatitis in humans and liver enlargement in rats and has similarity to idiosyncratic compounds such as felbamate. Kellye Daniels, Director of Toxicogenomics at Gene Logic, examined the gene expression data in more detail and found evidence for a transitory effect on Phase I and II drug-metabolizing enzymes, oxidative stress response genes, protein-repair associated genes, and a solute carrier. Taken together, this suggests the cells are metabolizing the conjugates for removal, proteins are damaged, and the cell is trying to remove them alongside a compensatory induction of a solute carrier to export them. These changes may reflect the reason rats do not develop obvious hepatotoxicity.

It is hypothesized that idiosyncratic compounds have a similar phenotypic pattern at the gene level to compounds known to induce hepatotoxicity in rats and that enables predictive models to detect the former as well as the latter. The effect of compounds on the expression of superoxide dismutase 2 (Sod2), shown in **Fig. 11**, will serve as an example. This gene encodes a protein that catalyzes the breakdown of superoxide into hydrogen peroxide and water in the mitochondria and is therefore a central regulator of reactive oxygen species (ROS) levels *(39)*. The expression of this gene is upregulated by compounds such as acetaminophen and lipopolysaccharide, known to induce

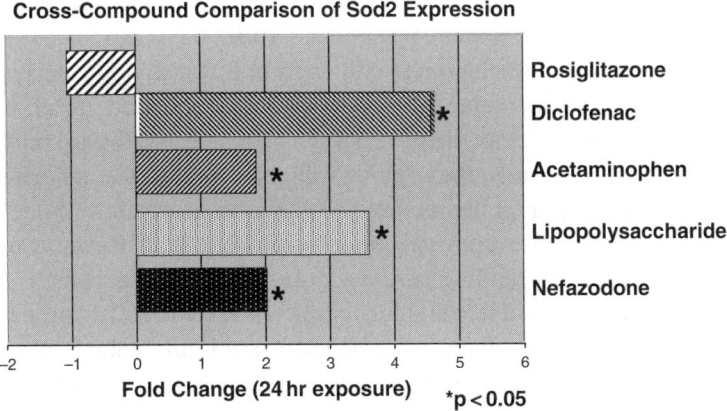

Fig. 11. The expression level of superoxide dismutase 2 (Sod2) in toxicant-treated liver tissue compared with vehicle controls is illustrated. Both the fold change and statistical results are shown.

hepatotoxicity in multiple species yet acting by differing mechanisms. Two idiosyncratic compounds, diclofenac and nefazodone, also elevate the level of this gene, whereas a drug not associated with hepatotoxicity, rosiglitazone, has no significant effect. This illustrates the commonality of gene expression among many compounds that cause human hepatotoxicity whether or not they induce similar toxicity in rats, the species examined here.

4. Use of Toxicogenomics in an *In Vitro* Approach

Because earlier identification of a compound destined to fail in the market-place would save significant resources, it is important to examine what can be done in terms of both cell culture (*in vitro*) and animal (*in vivo*) testing. Although cells in a culture environment do not maintain their *in situ* morphology, their homeostatic function and potentially the full complement of drug-metabolizing genes, the ease of use, and relative savings in terms of compound and time continues to induce researchers to exploit *in vitro* systems. Companies tend to focus on cell culture approaches to rank compounds on the basis of potential toxicity during lead optimization but only if suffi-cient throughput and cost constraints can be met *(3,40)*. Because the *in vitro* environment does not represent all of the complex processes at play upon *in vivo* exposure, during lead prioritization some prefer to focus on short-term *in vivo* approaches instead *(41)*.

How can toxicogenomics add value to classic *in vitro* approaches (e.g., cytotoxicity) and *in vivo* studies? First, it is helpful to identify what is needed in these stages. Donna Dambach and her colleagues at Bristol-Myers Squibb were working with a tiered testing strategy using immortalized human hepatocyte cell lines and primary rat and human hepatocytes *(40)*. They report good success in using their five human hepatocyte cell lines in a cytotoxicity assay wherein compounds with an IC_{50} value of <50 μM are deemed to be of increased risk and may be subjected to further testing. Although this assay only detects necrosis-causing compounds, they find it valuable because it has been reported that 50% of all drug-related hepatotoxicity is due to necrosis.

Because genomic approaches can monitor thousands of genes at one time, it can enlarge our understanding of the various cellular responses to an agent that can lead to both predictive models based on a statistical approach and to better mechanistic understanding. Barros and Martin (**Chapter 5** and Ref. *21*) detail a successful approach for using gene expression data from primary rat hepatocytes exposed to hepatotoxicants and safe compounds to generate a predictive toxicogenomic model of hepatotoxicity that can be used as a screen to rank compounds on the basis of their potential to induce hepatotoxicity. Similarly, Hultin-Rosenberg and colleagues at AstraZeneca report success in using such approaches *(22)*. These authors used several statistical methods and found little gene overlap between them. They concluded that the ability to "map biological pathways onto predictive sets will be problematic" as genes chosen with statistical approaches may or may not respond in similar ways to other members of the same pathway. Jeff Waring and his colleagues at Abbott have reported the use of toxicogenomics to identify mechanisms of toxicity using primary rat and human hepatocytes *(42,43)*, and Sawada et al. identified a set of genes in HepG2 cells that can be used in an *in vitro* screen to identify compounds that cause phospholipidosis *(44)*. Thus, teams of researchers are exploring new ways to improve accuracy of toxicity assessment using a genomics approach that may be used with *in vitro* or *in vivo* exposure to build predictive models and to enhance mechanistic understanding.

5. Regulatory Issues

The FDA has been proactive in the use of genomics and in March 2005 released a guidance paper on the use of pharmacogenomics (including toxicogenomics) in drug development. This document and recent presentations and papers by FDA staff can be found on their dedicated Web site (www.fda.gov/cder/genomics). **Chapter 3** and **Chapter 12** discuss in more detail issues and regulatory implications of genomic data. Some investigators have expressed concerns about the generalized use of toxicogenomics until factors such as quality control standards are in place, sufficient proof of concept studies emerge, and the concern about overinterpretation or misinterpretation of genomic data has been addressed *(8,11,45)*. Hopefully, some of these concerns have been alleviated by the recent spate of publications arising from the work of the MicroArray Quality Control (MAQC) project. (The articles can be obtained at the MAQC Web site www.fda.gov/nctr/science/centers/toxicoinformatics/maqc/.) This consortium reported good reproducibility of gene expression measurements between and within platforms using multiple platforms and test sites *(46,47)* and found analytical consistency across platforms when evaluating specific toxicogenomic studies *(48)*. The second phase of this work began in September 2006 and will provide a venue for individuals from government agencies (FDA, EPA, etc.), pharmaceutical companies, biotechnology companies, and platform providers to discuss differing approaches for the identification of biomarkers for preclinical and clinical use. Whereas some believe that the field of toxicogenomics is not "ready for prime time," others embrace its usefulness today in predictive, mechanistic, and biomarker applications particularly in the investigative and compound ranking aspects of drug development *(2,9,12)*. Besides the MAQC efforts, other public efforts are involved in standardizing genomic controls, analysis tools, protocols, and identifying minimal genomic data submissions requirements. These include the External RNA Controls Consortium (ERCC) and the HL7/CDISC/I3C Pharmacogenomics Data Standards Committee.

Even if the community can find common ground on data-mining approaches applied to genomic data, there remains a concern as to biomarker qualification particularly as a designation of *valid* confers mandatory submission of such data with the drug application. Although many use the term *validation*, Dr. Woodcock (Deputy Commissioner for Operations and Chief Operating Officer, FDA) suggested that *qualification* may be a better term as the former means many things to different people and tends to remind people of the physical test and not the underlying biological truth (Woodcock J, Presentation at the FDA, DIA, PhRMA, BIO, & PWG Workshop on Application and validation of genomic biomarkers for use in drug development and regulatory submissions. 2005. http://www.fda.gov/cder/genomics/presentations.htm). Some uneasiness exists in the pharmaceutical industry related to how biomarkers will be validated (i.e., qualified) and how they will learn of this as it does then elicit reporting requirements. As a beginning, the FDA recently published examples of valid genomic biomarkers and how they are affecting drug labels ("Table of Valid Genomic Biomarkers in the Context of Approved Drug Labels" found at www.fda.gov/cder/genomics/genomic biomarkers table.htm.). However, because no method exists today for the qualification of biomarkers, Goodsaid and Frueh in the Office of Clinical Pharmacology (CDER/FDA) have taken the initiative to propose a process *(49)*. There is an ongoing collaboration between the FDA and Novartis to identify a process map and examine biomarkers of renal injury in the rat with the goal of extending this work into man in a consortium to fully validate the biomarkers (Maurer, G, Presentation at the FDA, DIA, PhRMA, BIO, & PWG Workshop on Application and validation of genomic biomarkers for use in drug development and regulatory submissions. 2005. http://www.fda.gov/cder/genomics/presentations.htm). A consortium approach will further widespread acceptance of methods to qualify biomarkers as well as the biomarkers themselves. **Chapter 11** will address public consortia in more detail.

6. Conclusion

Toxicogenomics is posed to augment and improve the safety testing of compounds but not replace classic testing as some originally proposed. If applied appropriately with correct study designs, genomics can have widespread uses in toxicology. It can identify toxicity at doses of compounds that do not cause overt toxicity as shown by Heinloth and her colleagues at NIEHS

(18), speed detection of nongenotoxicity as reported by Peter Lord and co-workers at Johnson and Johnson *(19)*, detect rat-specific effects such as those induced by PPAR-α agonists *(15)*, and identify idiosyncratic compounds prior to human exposure using predictive models built on exposed primary rat hepatocytes or rat liver as reported by Hultin-Rosenberg and colleagues at AstraZeneca *(22)* and Martin et al. at Millennium Pharmaceuticals *(21)*. More and more companies are using this new technology in such predictive applications, mechanistic evaluations, and/or in biomarker discovery. The use of toxicogenomics and the balance between the three areas tend to be based on the number of compounds being evaluated among other issues. However, complicating the adoption of toxicogenomics is the interdisciplinary aspect of the field wherein critical analysis and understanding by people from many backgrounds including toxicologists, chemists, bioinformaticians, biostatisticians, pathologists, and regulators is needed. It is beneficial to have such individuals together in discussions of toxicogenomics so each can contribute their expertise to a facet of the field. Unfortunately, such groups rarely communicate effectively in large organizations although many pharmaceutical companies are beginning to realign teams of researchers through discovery and development to better meet the new demands of successful drug development. Hopefully, new technologies including toxicogenomics can be used to improve the prediction of drug toxicity prior to human exposure to provide better guidance in protecting human safety.

References

1. Kola, I. and Landis, J. (2004) Can the pharmaceutical industry reduce attrition rates? *Nat. Rev. Drug Discov.* **3**, 711–715.
2. Mayne, J.T., Ku, W.W., and Kennedy, S.P. (2006) Informed toxicity assessment in drug discovery: systems-based toxicology. *Curr. Opin. Drug Discov. Dev.* **9**, 75–83.
3. Barros, S.A. (2005) The importance of applying toxicogenomics to increase the efficiency of drug discovery. *Pharmacogenomics* **6**, 547–550.
4. DiMasi, J.A., Hansen, R.W., and Grabowski, H.G. (2003) The price of innovation: new estimates of drug development costs. *J. Health Econ.* **22**, 151–185.
5. Olson, H., Betton, G., Robinson, D., Thomas, K., Monro, A., Kolaja, G., et al. (2000) Concordance of the toxicity of pharmaceuticals in humans and in animals. *Regul. Toxicol. Pharmacol.* **32**, 56–67.
6. Suter, L., Babiss, L.E., and Wheeldon, E.B. (2004) Toxicogenomics in predictive toxicology in drug development. *Chem. Biol.* **11**, 161–171.
7. DiMasi, J.A. (2002) The value of improving the productivity of the drug development process: faster times and better decisions. *Pharmacoeconomics* **20** (Suppl 3), 1–10.
8. Boverhof, D.R. and Zacharewski, T.R. (2006) Toxicogenomics in risk assessment: applications and needs. *Toxicol. Sci.* **89**, 352–360.

9. Yang, Y., Blomme, E.A., and Waring, J.F. (2004) Toxicogenomics in drug discovery: from preclinical studies to clinical trials. *Chem. Biol. Interact.* **150**, 71–85.

10. Luhe, A., Suter, L., Ruepp, S., Singer, T., Weiser, T., and Albertini, S. (2005) Toxicogenomics in the pharmaceutical industry: hollow promises or real benefit? *Mutat. Res.* **575**, 102–115.

11. Lord, P.G. (2004) Progress in applying genomics in drug development. *Toxicol. Lett.* **149**, 371–375.

12. Searfoss, G.H., Ryan, T.P., and Jolly, R.A. (2005) The role of transcriptome analysis in pre-clinical toxicology. *Curr. Mol. Med.* **5**, 53–64.

13. Guerreiro, N., Staedtler, F., Grenet, O., Kehren, J., and Chibout, S.D. (2003) Toxicogenomics in drug development. *Toxicol. Pathol.* **31**, 471–479.

14. Dix, D.J., Houck, K.A., Martin, M.T., Richard, A.M., Setzer, R.W., and Kavlock, R.J. (2007) The ToxCast Program for Prioritizing Toxicity Testing of Environmental Chemicals. *Toxicol. Sci.* **95**, 5–12.

15. Lord, P.G., Nie, A., and McMillian, M. (2006) Application of genomics in preclinical drug safety evaluation. *Basic Clin. Pharmacol. Toxicol.* **98**, 537–546.

16. Shioda, T., Chesnes, J., Coser, K.R., Zou, L., Hur, J., Dean, K.L., et al. (2006) Importance of dosage standardization for interpreting transcriptomal signature profiles: evidence from studies of xenoestrogens. *Proc. Natl. Acad. Sci. U. S. A.* **103**, 12033–12038.

17. Farr, S. and Dunn, R.T. II (1999) Concise review: gene expression applied to toxicology. *Toxicol. Sci.* **50**, 1–9.

18. Heinloth, A.N., Irwin, R.D., Boorman, G.A., Nettesheim, P., Fannin, R.D., Sieber, S.O., et al. (2004) Gene expression profiling of rat livers reveals indicators of potential adverse effects. *Toxicol. Sci.* **80**, 193–202.

19. Nie, A.Y., McMillian, M., Brandon, P.J., Leone, A., Bryant, S., Yieh, L., et al. (2006) Predictive toxicogenomics approaches reveal underlying molecular mechanisms of nongenotoxic carcinogenicity. *Mol. Carcinog.* **45**, 914–933.

20. Mattes, W.B., Daniels, K.K., Summan, M., Xu, Z.A., and Mendrick, D.L. (2006) Tissue and species distribution of the glutathione pathway transcriptome. *Xenobiotica* **36**, 1081–1121.

21. Martin, R., Rose, D., Yu, K., and Barros, S. (2006) Toxicogenomics strategies for predicting drug toxicity. *Pharmacogenomics* **7**, 1003–1016.

22. Hultin-Rosenberg, L., Jagannathan, S., Nilsson, K.C., Matis, S.A., Sjogren, N., Huby, R.D., et al. (2006) Predictive models of hepatotoxicity using gene expression data from primary rat hepatocytes. *Xenobiotica* **36**, 1122–1139.

23. Kaplowitz, N. (2005) Idiosyncratic drug hepatotoxicity. *Nat. Rev. Drug Discov.* **4**, 489–499.

24. Dieckhaus, C.M., Miller, T.A., Sofia, R.D., and Macdonald, T.L. (2000) A mechanistic approach to understanding species differences in felbamate bioactivation: relevance to drug-induced idiosyncratic reactions. *Drug Metab. Dispos.* **28**, 814–822.

25. Alden, C., Lin, J., and Smith, P. (2003) Predicting toxicology technology for avoiding idiosyncratic liver injury. *Preclinica* **May/June**, 27–35.
26. Popovic, M., Nierkens, S., Pieters, R., and Uetrecht, J. (2004) Investigating the role of 2-phenylpropenal in felbamate-induced idiosyncratic drug reactions. *Chem. Res. Toxicol.* **17**, 1568–1576.
27. Ettinger, A.B., Jandorf, L., Berdia, A., Andriola, M.R., Krupp, L.B., and Weisbrot, D.M. (1996) Felbamate-induced headache. *Epilepsia* **37**, 503–505.
28. McGee, J.H., Erikson, D.J., Galbreath, C., Willigan, D.A., and Sofia, R.D. (1998) Acute, subchronic, and chronic toxicity studies with felbamate, 2-phenyl-1.,3-propanediol dicarbamate. *Toxicol. Sci.* **45**, 225–232.
29. Carvajal Garcia-Pando, A., Garcia, d.P., Sanchez, A.S., Velasco, M.A., Rueda de Castro, A.M., and Lucena, M.I. (2002) Hepatotoxicity associated with the new antidepressants. *J. Clin. Psychiatry* **63**, 135–137.
30. Stewart, D.E. (2002) Hepatic adverse reactions associated with nefazodone. *Can. J. Psychiatry* **47**, 375–377.
31. Choi, S. (2003) Nefazodone (Serzone) withdrawn because of hepatotoxicity. *CMAJ.* **169**, 1187.
32. Aranda-Michel, J., Koehler, A., Bejarano, P.A., Poulos, J.E., Luxon, B.A., Khan, C.M., et al. (1999) Nefazodone-induced liver failure: report of three cases. *Ann. Intern. Med.* **130**, 285–288.
33. Schrader, G.D. and Roberts-Thompson, I.C. (1999) Adverse effect of nefazodone: hepatitis. *Med. J. Aust.* **170**, 452.
34. Kostrubsky, S.E., Strom, S.C., Kalgutkar, A.S., Kulkarni, S., Atherton, J., Mireles, R., et al. (2006) Inhibition of hepatobiliary transport as a predictive method for clinical hepatotoxicity of nefazodone. *Toxicol. Sci.* **90**, 451–459.
35. Greene, D.S. and Barbhaiya, R.H. (1997) Clinical pharmacokinetics of nefazodone. *Clin. Pharmacokinet.* **33**, 260–275.
36. DeVane, C.L., Donovan, J.L., Liston, H.L., Markowitz, J.S., Cheng, K.T., Risch, S.C., and Willard, L. (2004) Comparative CYP3A4 inhibitory effects of venlafaxine, fluoxetine, sertraline, and nefazodone in healthy volunteers. *J. Clin. Psychopharmacol.* **24**, 4–10.
37. Kalgutkar, A.S., Vaz, A.D., Lame, M.E., Henne, K.R., Soglia, J., Zhao, S.X., et al. (2005) Bioactivation of the nontricyclic antidepressant nefazodone to a reactive quinone-imine species in human liver microsomes and recombinant cytochrome P450 3A4. *Drug Metab. Dispos.* **33**, 243–253.
38. Kostrubsky, V.E., Strom, S.C., Hanson, J., Urda, E., Rose, K., Burliegh, J., et al. (2003) Evaluation of hepatotoxic potential of drugs by inhibition of bile-acid transport in cultured primary human hepatocytes and intact rats. *Toxicol. Sci.* **76**, 220–228.
39. Landis, G.N. and Tower, J. (2005) Superoxide dismutase evolution and life span regulation. *Mech. Ageing Dev.* **126**, 365–379.
40. Dambach, D.M., Andrews, B.A., and Moulin, F. (2005) New technologies and screening strategies for hepatotoxicity: use of in vitro models. *Toxicol. Pathol.* **33**, 17–26.

41. Meador, V., Jordan, W., and Zimmermann, J. (2002) Increasing throughput in lead optimization in vivo toxicity screens. *Curr. Opin. Drug Discov. Dev.* **5**, 72–78.
42. Liguori, M.J., Anderson, L.M., Bukofzer, S., McKim, J., Pregenzer, J.F., Retief, J., et al. (2005) Microarray analysis in human hepatocytes suggests a mechanism for hepatotoxicity induced by trovafloxacin. *Hepatology* **41**, 177–186.
43. Waring, J.F., Ciurlionis, R., Jolly, R.A., Heindel, M., and Ulrich, R.G. (2001) Microarray analysis of hepatotoxins in vitro reveals a correlation between gene expression profiles and mechanisms of toxicity. *Toxicol. Lett.* **120**, 359–368.
44. Sawada, H., Takami, K., and Asahi, S. (2005) A toxicogenomic approach to drug-induced phospholipidosis: analysis of its induction mechanism and establishment of a novel in vitro screening system. *Toxicol. Sci.* **83**, 282–292.
45. Reynolds, V.L. (2005) Applications of emerging technologies in toxicology and safety assessment. *Int. J. Toxicol.* **24**, 135–137.
46. Shi, L., Reid, L.H., Jones, W.D., Shippy, R., Warrington, J.A., Baker, S.C., et al. (2006) The MicroArray Quality Control (MAQC) project shows inter- and intraplatform reproducibility of gene expression measurements. *Nat. Biotechnol.* **24**, 1151–1161.
47. Canales, R.D., Luo, Y., Willey, J.C., Austermiller, B., Barbacioru, C.C., Boysen, C., et al. (2006) Evaluation of DNA microarray results with quantitative gene expression platforms. *Nat. Biotechnol.* **24**, 1115–1122.
48. Guo, L., Lobenhofer, E.K., Wang, C., Shippy, R., Harris, S.C., Zhang, L., et al. (2006) Rat toxicogenomic study reveals analytical consistency across microarray platforms. *Nat. Biotechnol.* **24**, 1162–1169.
49. Goodsaid, F. and Frueh, F. (2006) Process map proposal for the validation of genomic biomarkers. *Pharmacogenomics* **7**, 773–782.

2

Use of Traditional End Points and Gene Dysregulation to Understand Mechanisms of Toxicity: Toxicogenomics in Mechanistic Toxicology

Wayne R. Buck, Jeffrey F. Waring, and Eric A. Blomme

Summary

Microarray technologies can be used to generate massive amounts of gene expression information as an initial step to decipher the molecular mechanisms of toxicologic changes. Identifying genes whose expression is associated with specific toxic end points is an initial step in predicting, characterizing, and understanding toxicity. Analysis of gene function and the chronology of gene expression changes represent additional methods to generate hypotheses of the mechanisms of toxicity. Follow-up experiments are typically required to confirm or refute hypotheses derived from toxicogenomic data. Understanding the mechanism of toxicity for a compound is a critical step in forming a rational plan for developing counterscreens for toxicity and for increasing productivity of research and development while decreasing the risk of late-stage failure in pharmaceutical development.

Key Words: gene expression; mechanism of toxicity; microarray; phenotype.

1. Introduction

A study published in 2002 demonstrated a robust, statistically significant link between wine drinking and healthy diet, nonsmoking, and higher household income (1). Yet, no one would presume that starting to drink wine would cause someone to suddenly choose a healthier lifestyle or get a pay increase! Whereas the distinction between cause and effect is clear in this simple example, analysis of cause and effect in the practice of toxicogenomics is less obvious. Statistical analysis of a large set of microarray gene expression profiles derived from

From: *Methods in Molecular Biology, vol. 460: Essential Concepts in Toxicogenomics*
Edited by: D. L. Mendrick and W. B. Mattes © Humana Press, Totowa, NJ

treatments with particular toxicants will essentially always produce a signature set of genes that can distinguish treatment from control groups. However, the predictive or diagnostic value of such a signature, as it applies to the practical toxicologic evaluation of compounds, requires validation as described in previous chapters. Nonetheless, gene expression signatures that discriminate based on epiphenomena instead of mechanisms of toxicity may fail to recognize toxic changes caused by new chemical series. Therefore, establishing gene expression signatures that are causally linked to a specific toxic process, so-called phenotypic anchoring, can increase one's confidence in the value of such signatures. Signatures can also be developed that will detect the activation of mechanistically important pathways in test animal species that are known to be associated with potential toxicity in humans but have no recognized phenotypic outcomes in test animals.

Toxicogenomic data are also extremely valuable to formulate new hypotheses regarding the mechanisms responsible for undesirable phenotypic changes. The extraordinary wealth of information on cell and tissue transcriptomes gained through microarrays can be combined with annotation libraries to group expression data into functionally related pathways, a process called pathway analysis *(2)*. Data collection at multiple time points can further imply cause-effect relationships between early and later gene expression. Investigation into mechanisms of toxicity is particularly helpful when a toxic phenotype, such as hepatocellular hypertrophy, can be caused by a number of distinct molecular mechanisms. In pharmaceutical discovery and development, efficient assays can then be designed to screen backup compounds against activation of the relevant pathway to select less toxic alternatives.

In this chapter, we will first discuss the situation where a single phenotype (liver hypertrophy) is associated with a number of distinct gene expression profiles, each of which has unique implications in safety assessment. Second, we will cover the utility of a cardiotoxic agent to pinpoint mechanisms of toxicity in a tissue having a limited range of morphologic and functional responses to injury, such as the heart. Third, the value of toxicogenomics in distinguishing beneficial, toxic, and incidental mechanisms of estrogens on the uterus will be explored as it relates to environmental and pharmaceutical toxicology. Fourth, we will highlight the advantage of using toxicogenomics to screen environmental compounds for testicular toxicity and the challenge of determining a mechanism of action in this complex tissue. Finally, we will discuss the value of toxicogenomics in discovering the activation of phenotypically inapparent cellular stress pathways as a predisposing mechanism for the development of idiosyncratic drug reactions and the ability to test for these changes in animal models.

2. Liver Hypertrophy: One Phenotype with Many Mechanisms

Gene expression profiles allow for a global, detailed evaluation of the molecular events that precede and accompany toxicity and therefore are extremely useful to understand the molecular mechanisms underlying various histopathologic or clinical chemistry changes. This is best illustrated with liver enlargement, a relatively frequent observation in rodents treated with xenobiotic agents. There are a variety of mechanisms by which drug-induced hepatomegaly can occur, including increases in smooth endoplasmic reticulum contents or in cytochrome P450 monooxygenase activities, peroxisome proliferation, hypertrophy of mitochondria, or hepatocellular proliferation (3). In some cases, these changes are adaptive and are not considered of toxicologic significance (4). In other cases, these changes are rodent specific and are not considered relevant to humans, such as peroxisome proliferation (5). In contrast, enzyme induction responses can sometimes be associated with liver injury, as evidenced by increased alanine aminotransferase (ALT) levels and hepatocellular apoptosis and necrosis (6,7). Finally, many compounds that induce liver hypertrophy fall into the category of nongenotoxic carcinogens (NGTCs). NGTCs are agents that do not cause a direct effect on DNA (i.e., nongenotoxic) yet test positive in long-term rodent carcinogenicity studies. Whereas the relevance of these findings to humans is not always clear, a positive result in this bioassay will result in a tremendous amount of time and resources allocated to identifying the mechanism and determining its relevance to humans (8). Thus, in the case of liver hypertrophy, toxicogenomics represents a useful addition to rapidly understand the underlying mechanism, which helps in interpreting and positioning liver findings.

In one of the pioneering toxicogenomic studies, Hamadeh et al. used gene expression profiling to distinguish different classes of enzyme inducers (9). In this study, rats were treated with three peroxisome proliferators (clofibrate, Wyeth 14,643, and gemfibrozil) and an enzyme inducer (phenobarbital). Phenobarbital is known to induce a variety of drug-metabolizing enzymes, including cytochrome P450 2B (CYP2B), 2C, and 3A (10). Rats treated for a single day with all compounds showed no histopathologic changes, whereas drug-related microscopic hepatocellular hypertrophy was observed in the livers of all animals after 2 weeks of treatment. Gene expression analysis conducted on the livers from the 24-h–treated rats revealed distinct gene expression patterns induced by the peroxisome proliferators compared with phenobarbital. Further refinement of the gene expression analysis identified a set of 22 genes that could accurately classify blinded liver hypertrophy–inducing compounds as being either compounds that induce peroxisome proliferation or phenobarbital-like compounds (9).

In our studies, we have used gene expression analysis to identify the mechanism underlying liver hypertrophy induced by an experimental anti-inflammatory agent, A-277249 *(6)*. In this study, treatment of rats for 3 days by A-277249 resulted in a twofold increase in liver weights, coupled with hepato-cellular hypertrophy. A toxicogenomic analysis on the livers from treated rats revealed that A-277749 induced gene expression changes highly similar to those induced by the aromatic hydrocarbon receptor agonists 3-methylcholanthrene and Aroclor 1254, including increases in CYP1A1, CYP1B1, and glutathione-*S*-transferase (**Fig. 1** and Color Plate 1). In addition, the gene expression analysis suggested that the induction of these enzymes might be driven by the targeted pharmacological activities of A-277249, making it unlikely that compounds in this class could be identified that would not result in liver hypertrophy. These studies allowed for a rapid termination of a program with little probability of success.

Toxicogenomics has also been applied to demonstrate that fumonisin mycotoxins cause liver hypertrophy and rodent liver carcinogenicity independent of the peroxisome proliferator activated receptor alpha (PPAR-α) pathway *(11)*. In this study, wild type and PPAR-α knockout mice were treated with fumonisin or Wy-14,463, a prototypical PPAR-α agonist. The gene expression analysis showed that PPAR-α was necessary for the regulation of

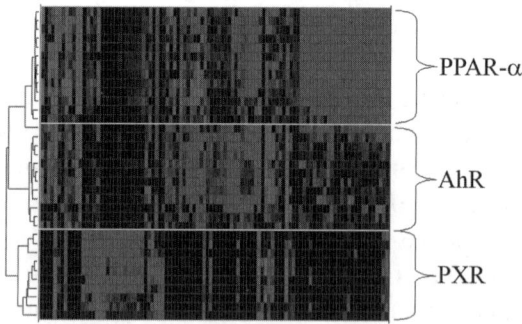

Fig. 1. Identification of genes regulated in the liver of rats after xenobiotic activation of the nuclear receptors PPAR-α, aromatic hydrocarbon receptor (AhR), or pregnane X receptor (PXR). The heatmap shows the genes significantly regulated in liver by several prototypical inducers of the three nuclear receptors. Genes shown in light gray are either up- or down-regulated relative to vehicle-treated control, and genes shown in black are unchanged. These data were extracted from the Iconix DrugMatrix database. (*see* Color Plate 1).

lipid metabolism genes associated with liver hypertrophy in Wy-14,463–treated mice, whereas treatment with fumonisin resulted in similar expression changes for the lipid metabolism genes in both wild type and PPAR-α knockout mice. These results confirmed that mouse liver carcinogenicity induced by fumonisin was independent of PPAR-α, thus suggesting that the toxicity induced by fumonisin is not necessarily species-specific in contrast with what is seen with PPAR-α agonists.

A number of studies have attempted to apply toxicogenomics in order to identify compounds associated with liver hypertrophy that are likely to test positive in long-term rodent carcinogenicity studies. Identifying these compounds early would result in significant savings in time and resources and would allow for the prioritization of compounds unlikely to test positive in long-term carcinogenicity studies. Fielden et al. *(12)* used gene expression data from rats treated for 5 days with more than 100 structurally and mechanistically diverse nongenotoxic hepatocarcinogens and non-hepatocarcinogens. From these data, the authors identified a gene expression signature consisting of 37 genes that could classify hepatocarcinogens and non-hepatocarcinogens with 86% and 81% sensitivity and specificity, respectively. In addition, by comparing the gene expression data with a reference database, this signature can provide an understanding of potential modes of action for hepatic tumorigenicity such as regenerative proliferation, proliferation associated with xenobiotic receptor activation, peroxisome proliferation, and steroid hormone–mediated mechanisms.

A similar approach was taken by Nie et al. *(13)*. In this study, male rats were treated for 1 day with 52 compounds, 24 of which were NGTCs and 28 were non-carcinogens. Microarray analysis on the livers from treated rats revealed a set of six genes that could distinguish NGTCs from non-carcinogenic compounds. Further work is under way to verify the robustness of these signatures across different laboratories and gene expression platforms. Nonetheless, these early proof-of-concept studies suggest that gene expression analysis represents a valid approach toward distinguishing NGTCs from non-carcinogens using short-term rodent studies. Overall, the toxicologic relevance of rodent liver hypertrophy to humans is variable and highly dependent on the underlying mechanism. Without an understanding of the mechanism, potentially safe and efficacious compounds may be deprioritized for further development. Thus, the application of gene expression analysis toward identifying the mechanism underlying drug-induced liver hypertrophy represents a clear example of how toxicogenomics can complement traditional toxicologic end points and bring value as a safety evaluation tool.

3. Cardiotoxicity: Use of a Tool Compound to Dissect the Mechanism of Toxicity

The heart is a relatively common target organ of toxicity for pharmaceutical, natural, or industrial agents. For pharmaceutical agents, hepatic toxicity is probably the most common toxicity identified, but cardiac toxicity is sufficiently prevalent to warrant interest by toxicologists. For the toxicologist in development, pharmaceutical agents can be withdrawn from the market because of cardiovascular toxicity or development compounds can be terminated because of the discovery of unanticipated cardiac toxic changes in chronic toxicology studies. For the discovery toxicologist, the challenge may be in selecting development candidates that will not induce development-limiting cardiovascular toxicology. Current systems to monitor functional deficits of the cardiovascular system are robust enough to ensure the safety of patients involved in clinical trials or prescribed agents with potential cardiac effects. In contrast, there is a lack of robust biomarkers to detect, assess, and monitor the progression of cardiac structural damage or altered cellular homeostasis *(14)*. Therefore, there is a significant interest in avoiding compounds that may affect myocardiocyte homeostasis or cause structural damage.

Histologic changes induced by cardiac toxicants are relatively limited in nature and usually are characterized by myocardiocyte degeneration, apoptosis, or necrosis. In addition, cardiac toxicity may be identified in subacute to chronic studies, at a time when the toxicity is well advanced and when it is difficult to formulate hypotheses regarding mechanisms. Ultrastructural evaluation may allow for an increased definition of the morphologic changes. For instance, mitochondrial swelling in degenerating cells is considered a good indication that mitochondria represent a primary target organelle of toxicity. However, for the most part, morphologic changes observed by histopathology or electron microscopy reveal no real clue regarding mechanism of cardiac toxicity. Gene expression profiling represents a unique approach to rapidly generate a vast amount of molecular data relevant to the toxic event and to use these data to formulate hypotheses that can then be confirmed or refuted in follow-up experiments. This concept will be illustrated in this section using literature examples but also a specific case example that we have encountered in our internal discovery programs.

Doxorubicin represents the ideal tool compound to interrogate cardiac toxicity, and consequently doxorubicin has been extensively used in the literature for biochemical studies of cardiac toxicity. Doxorubicin is an anthracycline antibiotic with broad-spectrum chemotherapeutic activity. The use of this agent is however limited by its cardiotoxicity, which may occur months to years after treatment *(15)*. The toxicity also occurs in animals, providing excellent preclinical models to understand the mechanism of cardiotoxicity of the anthracycline antibiotics. One of the proposed mechanisms of cardiotoxicity involves the formation of reactive oxygen species (ROS) via redox cycling of the semiquinone radical intermediate *(16,17)*. ROS formation results in damage to cellular macromolecules, such as DNA, proteins, or lipids *(17)*. In particular, cardiac mitochondria are considered the principal early organelle of toxicity due to damage by the ROS or disturbances in mitochondrial homeostasis *(16)*.

The cardiotoxicity of doxorubicin is difficult to detect after a single dose in the rat with serum chemistry and light microscopy *(18)*. In contrast, large numbers of relevant gene expression changes can be detected in the heart shortly after treatment with doxorubicin, and these early transcriptomic effects are useful to unravel the mechanism of toxicity. In various internal gene profiling studies, we have demonstrated that doxorubicin affects biological pathways related to mitochondrial function and calcium regulation, thereby demonstrating that gene expression profiles represent an ideal tool for hypothesis generation. Interestingly, these transcriptomic changes indicative of mitochondrial dysfunction occurred well before decreased ATP production, which can be shown using isolated mitochondria from the heart of doxorubicin-treated rats *(19,20)*. This reinforces the concept that gene expression changes most relevant to the mechanism of toxicity can occur well before actual organelle dysfunction. This property is very useful and can represent a very effective approach to select exploratory compounds without cardiac toxicity liabilities. This is best illustrated by a recent study that our discovery organization published *(21)*.

Our team was interested in developing inhibitors of acetyl-coA carboxylase (ACC) as potential therapeutic agents for hyperlipidemia. Evaluation of the early series revealed neurologic and cardiovascular liabilities in preclinical models, precluding further evaluation of this chemotype. This prompted us to investigate the mechanism of toxicity of this series, in an effort to develop a counterscreen that could be used in the selection of backup compounds. Briefly, male Sprague-Dawley rats were treated daily with toxic doses of the exploratory compounds for 3 days. Doxorubicin was used as a positive control in this study. As expected, no evidence of cardiotoxicity was evident after this short-term dosing period when evaluating histopathologic sections and serum chemistry panels. In contrast, the gene expression profiles induced in the heart by these compounds were very similar to those induced by doxorubicin, confirming their cardiotoxic liability and also suggesting similar mechanisms of toxicity. DrugMatrix is a commercial database that contains gene expression profiles of multiple organs of rats treated with a wide variety of pharmaceutical and toxicologic agents *(22)*. When these gene expression changes were compared with those present in the DrugMatrix commercial database, the expression profiles from the ACC inhibitors had strong correlations with a number of reference expression profiles, especially profiles induced by cardiotoxicants or cardiotonic agents, such as cyclosporin A, haloperidol, or norepinephrine, again suggesting cardiotoxic liability. Finally, when the gene expression changes were analyzed in the context of biological pathways, the mitochondrial oxidative phosphorylation pathway was mostly affected. Overall, these results indicated that this series of compounds had cardiotoxic liability due to a mechanism very similar to that of doxorubicin. This rapid mechanistic understanding allowed us to rapidly select safer chemotypes for this program by ensuring the lack of similar transcriptomic effects in the hearts of treated rats.

4. Estrogenic Activity: Complex Mechanisms Arising from Similarly Acting Compounds

Compounds with hormone-mimetic effects have the potential to influence fetal and prepubertal development, fertility, and mammary and reproductive carcinogenesis. The significance for carcinogenesis in high-dose screening studies depends, in part, on the particular hormone pathways that are activated or inhibited *(23)*. The contribution of toxicogenomics for understanding the mechanisms of action and toxicity for individual estrogenic compounds screened in the rat uterotropic assay is discussed after a brief review of the complexities of estrogen signaling.

Estrogen effects are mediated by nuclear-mediated and membrane-mediated mechanisms. Nuclear-mediated estrogen effects employ the estrogen receptor alpha (ER-α) and, recently discovered in 1996, the estrogen receptor beta (ER-β) *(24)*. ER-β and ER-α are encoded by separate genes with unique developmental and tissue distributions and each undergoes alternative splicing to form multiple isoforms *(25,26)*. In the presence of estradiol, the nuclear estrogen receptors undergo dimerization and activate transcription by binding to the estrogen response element. ER-β recruitment of transcription initiation factors is less efficient than that of ER-α, and these two estrogen receptors tend to have opposing effects within a particular cell *(26)*. The ER-α isoform ER-α 46 has a truncated N-terminus and is unable to bind to the estrogen response element, but when palmitoylated, it can associate with the Shc adapter protein and the insulin-like growth factor I receptor (IGF-IR) to transduce signals from the plasma membrane upon binding to estrogen *(27)*. Membrane-mediated estrogen effects are responsible for fast signaling events that have been described in vascular endothelium, neurons, and the myometrium. Direct phosphorylation of estrogen receptors by mitogen-activated protein kinase (MAPK) and growth factor receptors alters ER binding to coactivators and corepressors causing altered function in nuclear-mediated pathways *(28)*. Additionally, an orphan G-protein–coupled receptor, GPR30, is estrogen responsive independent of nuclear receptors through activation of adenylyl cyclase *(29)*. The convergence of membrane and nuclear estrogen signaling results in the modulation of transcription and cell function beyond those genes with estrogen response elements in their promoters *(30,31)*. Nuclear estrogen receptors undergo conformational changes when bound to ligand that reveal binding sites for coactivators and corepressors that have chromatin remodeling activity *(28)*. The coactivators are targets for modulation by signal transduction cascades themselves and are shared with other nuclear receptors leading to cross-talk between pathways *(32,33)*. ER tethering to other transcription factors, such as HIF-1 and Sp1, also allows estrogen signaling to modulate genes lacking an estrogen response element *(34,35)*. Finally, xenoestrogens and selective estrogen receptor modifiers (SERMs) each have distinct profiles of ER binding and recruitment of cofactors leading to diverse transcriptional responses *(28)*. The complexity of estrogen signaling makes predictive modeling of tissue-specific gene expression difficult, but a genomics approach linking gene expression to mechanisms of action is proving useful for evaluating risks in both environmental and pharmaceutical toxicology.

The toxicogenomics approach to linking gene expression and mechanism of action can be thought of in a stepwise manner. First, compounds are administered and gene expression profiles are collected from the tissue of interest. For example, the role of estradiol in rat uterus gene expression was studied by Wu and colleagues *(36)*. Estradiol was found to be necessary and sufficient for inducing the expression of a number of genes. Second, classification of estrogen-responsive genes according to annotations in gene databases using a predefined vocabulary (http://www.geneontology.org) allows speculative association of gene changes with observed phenotypic changes *(37)*. For example, Naciff and colleagues examined gene expression changes in prepubertal rat uterus after treatment with 17α-ethynyl estradiol, an estrogenic drug with estradiol-like effects *(38)*. They associated phenotypic changes of ethynyl estradiol–treated uterus (such as edema and hyperemia) with genes known to effect vascular function (e.g., vascular endothelial growth factor [VEGF] and cysteine-rich protein 61). Third, these gene-phenotype associations can be experimentally directly tested. For example, Heryanto and colleagues administered estradiol in ovariectomized rats and analyzed blood vessel and stromal cell density in the presence and absence of an antibody to VEGF *(39)*. The antibody caused increased stromal cell density, which was interpreted as having blocked uterine edema. Thus, the observation of uterine edema is *phenotypically anchored* to VEGF gene expression. Fourth, experimentally proven gene-phenotype associations can be used for screening future agents for undesirable toxic effects when the phenotype is below the sensitivity of detection or does not present itself in the model system used.

Toxicogenomics can enhance the sensitivity of screening assays for pharmacological effect. For example, phytoestrogens are present in laboratory animal feed containing soy, but it is unclear whether or not they interfere with the uterotropic assay for screening potential endocrine disruptors for estrogenic effects. Naciff and colleagues *(40)* found gene expression analysis to be more sensitive to estrogenic compound administration than phenotypic markers (i.e., wet uterus weight). Therefore, they performed gene expression analysis on prepubertal rats fed a phytoestrogen-containing diet compared with rats treated with a phenotypically subthreshold dose of ethynyl estradiol *(41)*. None of the animals had detectable changes in uterine wet weight. Gene expression was altered on the phytoestrogen-containing diet, but there were no genes in common between the subthreshold estrogen control and the phytoestrogen diet. The conclusion was that phytoestrogen-containing feed does not interfere with uterotropic screening assays. As a second example, the objective of a pharmaceutical project was to develop a new SERM with significantly reduced incidence of pelvic organ prolapse, which is a side effect of estrogen therapy, while preserving the beneficial antiproliferative effects in breast and uterine tissue and osteoprotective effects. Unfortunately, the rat uterotropic assay does not distinguish SERMs that predispose to pelvic organ prolapse. Helvering and colleagues have associated increased matrix metalloproteinase 2 (MMP2) gene expression with a greater risk of prolapse in humans and then translated this finding to comparative microarray analysis of rat uterine gene expression using multiple SERMs *(42)*. If MMP2 is proven to be involved in the mechanisms leading to pelvic organ prolapse, it could be exploited as a reliable indicator against which to screen compounds in laboratory animal species for the relative risk of this complication in humans.

5. Testicular Toxicity: Defining Toxic Mechanisms in a Complex Tissue

Compared with studies in the liver or kidney, there have been few published studies using gene expression profiling to investigate mechanisms of testicular toxicity. Yet, testicular toxicity is not uncommon during the development of pharmaceutical agents and can represent a challenge, as it may not be detected morphologically until chronic studies are conducted, but also because early histopathologic changes can be very subtle and can easily be missed unless more sophisticated techniques, such as tubular staging, are used *(43)*. Current biomarkers of toxicity (such as serum follicular stimulating hormone or semen analysis) are not robust enough to detect early changes both in preclinical studies and in clinical trials. A recent working group sponsored by the Institutional Life Sciences Institute (ILSI) Health and Environmental Sciences Institute

(HESI) has tried to address this issue through an evaluation of additional biomarkers, such as Inhibin B *(44)*. Data released do not suggest that plasma inhibin B represents a marker sufficiently sensitive to detect modest testicular dysfunction in rats. Because sensitive biomarkers are currently lacking, gene expression profiling represents a potentially useful approach to develop an improved understanding of the various mechanisms of testicular toxicity. This improved understanding could ultimately be used to discover new sensitive markers of testicular injury. This is of particular interest to the pharmaceutical industry, but especially in the field of environmental toxicology, as a significant number of pesticides and other environmental toxicants have known testicular effects.

Investigating the mechanism of testicular toxicity can be quite challenging, given the complexity of the testis as a tissue, but also the complexity of the regulatory mechanisms for testicular function. The testis is composed of several different cell types characterized by unique functional and morphologic features but having close paracrine interactions and complex cellular interdependence. *In vitro* models have been used to circumvent this complexity, but ultimately these models do not reflect the paracrine interactions occurring in the tissue. Furthermore, cell lines of Sertoli or Leydig cell origin have lost major functional characteristics, and primary cultures of testicular cells require significant resources of time and can only be used for short-term experiments. For instance, Leydig cell cultures are frequently used to interrogate specific mechanisms of toxicity. Cultures are prepared with Percoll gradient centrifugation methods but are usually not pure, and preparations only yield a limited number of cells that need to be used shortly after preparations. These primary cultures are, however, extremely useful to confirm or refute hypotheses. For instance, adult Leydig cells were used to further understand the mechanism by which Aroclor 1254, a polychlorinated biphenyl, disrupts gonadal function *(45)*. By reverse transcription PCR, the authors demonstrated treatment-induced downregulation of the transcripts of various enzymes involved in steroidogenesis. These results coupled with a demonstration of decreased basal and luteinizing hormone (LH)-stimulated testosterone and estradiol production confirmed that Leydig cells represent a primary target cell for Aroclor 1254 and that the mechanism of toxicity was partly related to an effect of steroidogenesis. A similar approach will be illustrated in a subsequent example.

Several studies have used gene expression profiling to investigate the molecular basis of testicular toxicity *(46–51)*. For instance, testicular gene expression profiles were generated using a custom nylon DNA array and evaluated after exposure of mice to bromochloroacetic acid, a known testicular toxicant. Transcript changes were detected involving genes with known functions in fertility, such as Hsp70-2 and SP22, as well as genes encoding proteins involved in cell communication, adhesion, and signaling, supporting the hypothesis that the toxicologic effect was the result of disruption of cellular interactions between Sertoli cells and spermatids *(49,52)*. Several studies have investigated using genomics technologies to identify the mechanisms of testicular toxicity associated with exposure to di(*n*-butyl) phthalate (DBP) *(50,53)*. DBP is a plasticizer widely used in products such as food wraps, blood bags, and cosmetics *(54)*. Because of its wide use, levels of DBP metabolites are detectable in human urine *(55)*. DBP is a male reproductive toxicant, and the developing testis is a primary target through a suspected antiandrogenic effect *(56)*. Using fetal rat testes exposed *in utero* to DBP, Shultz et al. evaluated global changes in gene expression and demonstrated reduced expression of several steroidogenic enzymes, thereby providing a molecular mechanism for the antiandrogenic effect of this agent *(50)*. These findings are nicely supported by and explained at the molecular level by other investigations showing that DPB impairs cholesterol transport and steroidogenesis in the fetal rat testis *(57)*. Likewise, a recent study investigated the mechanisms of toxicity associated with the triazole fungicides *(51)*. It is known that these agents can modulate in mammalian species many cytochrome mixed function oxidase (CYP) enzymes that are involved in the metabolism of xenobiotics and endogenous molecules. Testicular toxicity associated with these agents is typically not detected until chronic or reproductive studies are conducted. Using custom oligonucleotide microarrays, the authors confirmed that these agents altered the expression of many CYPs and affected the sterol and steroid metabolic pathways in the testis. Although these studies are preliminary, they do suggest several plausible mechanisms of toxicity that need to be further interrogated with specifically designed follow-up experiments.

In the pharmaceutical industry, the primary objective of gene expression profiling studies in the testis is to develop approaches that would allow one to weed out potential testicular toxicants at an early stage in the candidate selection process in order to avoid compound termination at an advanced stage of development. A recent study used this approach *(58)*. Using four prototypical testicular toxicants, this study evaluated whether gene expression profiling could facilitate the identification of compounds with testicular toxicity liabilities. In this study, gene expression profiles were generated 6 h after dosing. Not surprisingly, histopathologic changes were absent, except for one agent. However, the microarray analysis identified several differentially expressed genes that were consistent with the known action of the toxicants. Our laboratory has also evaluated this approach. However, in contrast with the above studies, our gene expression profiles were generated 1 and 4 days after treatment with the prototypical toxicants. As expected, these treatments did not result in significant histopathologic changes, although all toxicants used in the study were known to be potent testicular toxicants. In general, the 1-day time point did not prove very useful to generate robust and consistent expression profiles (**Fig. 2** and Color Plate 2). This is also our experience with other tissues where short-term dosing paradigms are associated with significant interindividual variability. We speculate that this time point is too early to reach a steady state in tissue exposure for most agents. In contrast, the 4-day time point was associated with robust and reproducible gene expression changes, although the number of differentially expressed genes was much lower than that seen in other tissues, such as liver or kidney. The reference compounds were selected to cover several mechanisms of toxicity, and as expected in the absence of histologic changes, there were few consistent genes differentially expressed across all treatment groups. However, these expression profiles were extremely useful to formulate hypotheses regarding mechanisms of toxicity. For instance,

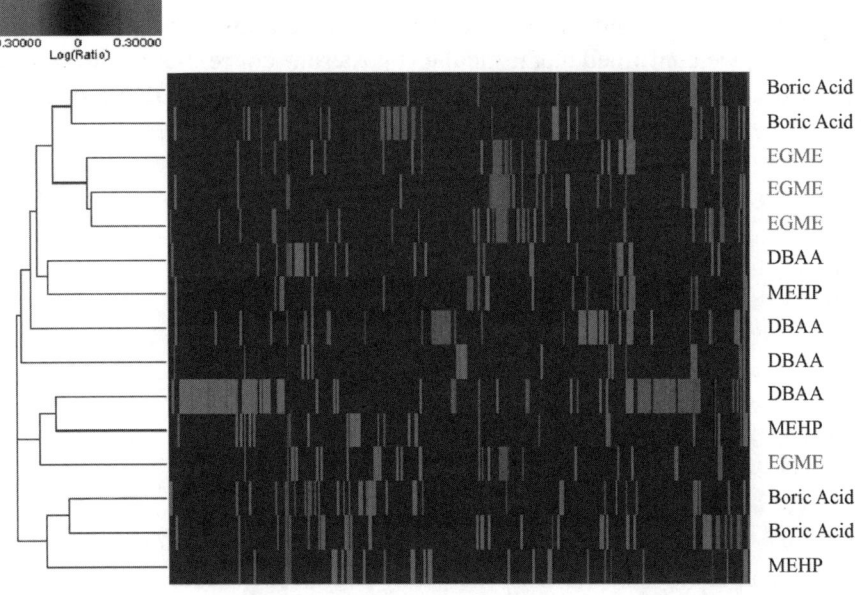

Fig. 2. Hierarchical clustering of gene expression profiles of the testes of male Sprague-Dawley rats treated with a single dose of various testicular toxicants and sacrificed 24 h after treatment. Included in the heatmap are genes that were upregulated or downregulated by a factor of ≥ 1.5 with a p value less than .01. Light gray indicates up- or down-regulation. In general, the testicular toxicants induced limited numbers of consistent gene expression changes after 1 day of treatment, suggesting that the treatment period was too short to generate a list of differentially expressed genes that could be used for mechanistic understanding. EGME, ethylene glycol monoethyl ether; DBAA, dibromoacetic acid; MEHP, mono-(2-ethylhexyl) phthalate. (*see* Color Plate 2).

two of the compounds used were the halogenated acetic acids dibromoacetic acid (DBAA) and dichloroacetic acid (DCAA). Daily oral treatment of male rats with DBAA and DCAA at high doses (250 mg kg^{-1} day^{-1} and 500 mg kg^{-1} day^{-1}, respectively) induces specific early morphologic changes in the testis, characterized by failed spermiation or failure of release by Sertoli cells of mature step 19 spermatids. Whereas this morphologic change strongly suggests that the Sertoli cell is the target cell of DBAA and DCAA, our toxicogenomics evaluation using whole testes indicated that both compounds induced a consistent downregulation of cytochrome P450c17α (CYP17). CYP17 is expressed in the testis by Leydig cells and represents a key enzyme for the production of gonadal testosterone by catalyzing the conversion of progesterone to hydroxyprogesterone and androstenedione *(59,60)*. Consequently, decreased levels of this enzyme would lead to decreased testosterone production by Leydig cells, suggesting that halogenated acetic acids induce testicular toxicity partly through an effect on testicular testosterone production. Decreased CYP17 mRNA levels were confirmed by real-time reverse transcription PCR (**Fig. 3**). In addition, we confirmed that testicular testosterone concentrations were significantly decreased in the testis of rats treated for 4 days with DBAA or DCAA, thereby demonstrating the biological implications of the decrease in CYP17 transcript levels.

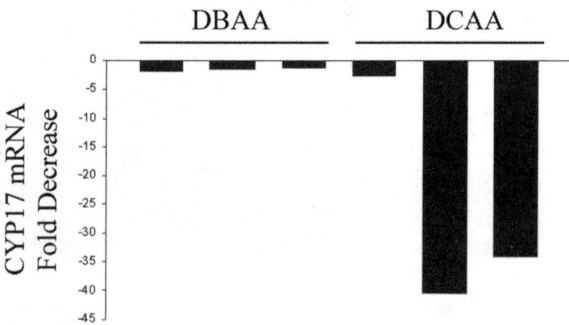

Fig. 3. Quantification of CYP17 mRNA levels by real-time quantitative reverse transcription (Taqman) PCR. To confirm the downregulation of CYP17 mRNA in the testes of rats treated for 4 days with either DBAA or DCAA, mRNA levels of CYP17 were quantified with RT-PCR and normalized to the average of three control rats for each compound. Note that whereas DBAA induces a consistent twofold decrease in CYP17 mRNA levels, DCAA treatment was associated with a significant decrease of CYP17 mRNA levels.

6. Idiosyncratic Drug Reactions: Mechanisms Without a Phenotype

There are times when discrepancies may exist between gene expression analysis and classic phenotype or functional measurements of toxicity or when gene expression profiles may be used to detect or understand changes that may never manifest at the phenotypic level. This is best illustrated by drugs that induce idiosyncratic drug reactions. Idiosyncratic drug reactions (IDRs) are defined as adverse drug reactions that occur in a small minority of exposed people and are not related to the pharmacology of the drug *(61)*. The mechanism underlying the toxicity is unknown, however several theories exist regarding how IDRs occur, including drug-protein adduct formation, mitochondrial toxicity, inflammatory reactions, and genetic polymorphisms *(62,63)*. This type of toxicity is not limited to a specific class of drugs and has been observed with a wide spectrum of therapeutic classes including antibacterials, insulin-sensitizing drugs, and antiepileptics. Idiosyncratic toxicity reactions often manifest in the liver; however, other tissues can be targets as well, such as the heart and skin *(64)*.

Idiosyncratic adverse effects that have been observed in humans are often not detected in preclinical models using traditional toxicology end points, thus making it difficult to predict these types of reactions before clinical trials or market launch. This could be due to the fact that the adverse event is species-specific; alternatively, because these types of adverse events are rare with a typical occurrence of 1 in 100 to 1 in 100,000 patients, typical adverse outcomes are most likely statistically impossible to detect in traditional preclinical safety studies *(62)*. However, despite the lack of significant changes in histopathologic or clinical chemistry parameters associated with IDR compounds in preclinical species, in some cases it may be possible to identify gene expression changes induced by the compound that are associated with toxicity.

Our laboratory has had interest in further understanding IDR at the molecular level. For instance, an internal compound (Cpd-001) was tested in multiple preclinical species and did not display overt signs of hepatotoxicity. However, upon reaching the clinic, several subjects did experience elevated levels of ALT. In order to see if gene expression changes related to hepatotoxicity could be observed in a preclinical species, we treated rats with high oral doses of Cpd-001 for 3 days. As had been reported previously, there were no significant test article–related changes in the clinical chemistry or hematology parameters evaluated, except for slight increases in serum ALT and aspartate aminotransferase (AST) that could not be considered toxicologically significant. In contrast, gene expression analysis of the livers from rats treated with Cpd-001 revealed a large number of gene expression changes. The gene expression changes were compared with expression profiles from reference compounds in DrugMatrix *(65)*. The hepatic expression profiles induced by Cpd-001 treatment

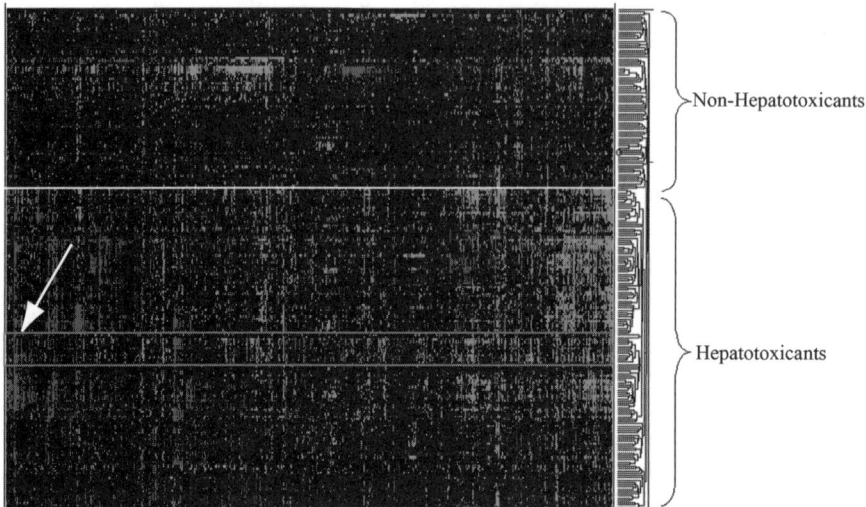

Fig. 4. Heatmap of gene expression profiles from the liver of rats treated with Cpd-001 (arrow) and a wide variety of reference compounds including nonhepatotoxicants and hepatotoxicants. The expression profiles induced by Cpd-001 are highly similar to and cluster with the expression profiles from the reference hepatotoxicants in the Iconix DrugMatrix database. Light gray indicates up- or down-regulation. (*see* Color Plate 3).

showed a significant correlation with expression profiles induced in the liver by a number of known hepatotoxicants (**Fig. 4** and Color Plate 3). In addition, gene expression changes indicative of an oxidative stress response were observed after treatment with Cpd-001. Thus, in this example, there were a significant number of gene expression changes suggestive of hepatotoxicity despite the lack of histopathologic or clinical chemistry changes of toxicologic significance.

Similar findings were reported for the drug felbamate *(66)*. Felbamate is an antiepileptic drug that shows only minimal toxicity in preclinical species but produces devastating idiosyncratic toxicity in a small percentage of patients characterized by irreversible aplastic anemia and fulminant liver failure *(67)*. The IDR associated with felbamate is thought to be the result of reactive metabolite formation. In this study, gene expression analysis was conducted on the liver of rats treated with felbamate. Although felbamate did not induce overt hepatotoxicity in rats, gene expression changes highly associated with oxidative stress and reactive metabolite formation could be detected. Altogether, these examples suggest that gene expression analysis may be more sensitive than traditional functional end points. However, for the most part, these cases are exceptions.

There are other instances where gene expression analysis and traditional measurements of toxicity may differ, such as when clinical chemistry or histopathologic analyses identify changes not readily apparent with gene expression analysis. These situations may be explained by the fact that the gene expression changes may have occurred before the morphologic manifestation of toxicity and may have subsequently normalized and returned to baseline. In other cases, a lack of gene expression changes may be due to the fact that transcriptional changes occurred in only a small subset of cells. This is especially true for the complex and heterogeneous tissues, where techniques such as laser capture microdissection may be needed to identify gene expression changes limited to a small cellular compartment. For example, the transcriptome of esophageal epithelium was specifically evaluated in rats in a model of esophagitis *(68)*. Because the initial target of cell injury is the epithelium, microdissection can reveal important gene expression changes in the target cell type, which only comprises a small amount of the total tissue, and eliminates downstream expression changes occurring in nearby tissues.

7. Conclusion

As illustrated in preceding examples, gene expression data represent a unique source for hypothesizing mechanisms of toxicity based on an exhaustive screening of tissue functions. Gene expression profiling can distinguish between multiple molecular mechanisms that result in a final common phenotype, such as liver hypertrophy. In some cases, the limited morphologic repertoire of a tissue's response to injury, such as in the heart, can be further refined with transcriptomics and the use of prototypical toxicants to probe for mechanisms associated with toxicity. Estrogenic compounds are exemplary of the diverse mechanisms and phenotypic responses that can be stimulated from a single compound or class of compounds and the necessity to evaluate which mechanisms are relevant for particular toxicology concerns. The challenges of determining mechanisms of action in a complex tissue like the testis, and the ability to discover expression profiles that predict increased susceptibility to rare idiosyncratic reactions where a phenotype is usually absent, highlight the unique contribution of toxicogenomics to pharmaceutical and environmental toxicology. Moving from statistical association of gene expression and phenotype toward proven mechanistic linkage avoids dependence upon signatures based on epiphenomena and allows the development of more efficient screening assays for specific toxic effects. The cellular target of toxicity may respond to compound administration by a number of pathways that may be adaptive or of no consequence to the health of the organism. Mechanistic clarity aids in the sharpening of expression signatures by excluding these irrelevant pathways and inviting the study of relevant homologous pathways in other test species. Thus, exploration of the mechanisms of toxicity stands to greatly enhance the reliability of gene signatures used to predict activation of specific pathways of toxicity. This, in turn, should result in an improved hazard identification and a more robust risk assessment.

References

1. Barefoot, J. C., Gronbaek, M., Feaganes, J. R., McPherson, R. S., Williams, R. B., and Siegler, I. C. (2002) Alcoholic beverage preference, diet, and health habits in the UNC Alumni Heart Study. *Am. J. Clin. Nutr.* **76**, 466–472.
2. Bono, H., and Okazaki, Y. (2002) Functional transcriptomes: comparative analysis of biological pathways and processes in eukaryotes to infer genetic networks among transcripts. *Curr. Opin. Struct. Biol.* **12**, 355–361.
3. Larrey, D., Tinel, M., and Pessayre, D. (1983) Formation of inactive cytochrome P-450 Fe(II)-metabolite complexes with several erythromycin derivatives but not with josamycin and midecamycin in rats. *Biochem. Pharmacol.* **32**, 1487–1493.
4. Amacher, D. E., Schomaker, S. J., Boldt, S. E., and Mirsky, M. (2006) The relationship among microsomal enzyme induction, liver weight, and histological change in cynomolgus monkey toxicology studies. *Food Chem. Toxicol.* **44**, 528–537.
5. Lai, D. Y. (2004) Rodent carcinogenicity of peroxisome proliferators and issues on human relevance. *J. Environ. Sci. Health C Environ. Carcinog. Ecotoxicol. Rev.* **22**, 37–55.
6. Waring, J. F., Gum, R., Morfitt, D., Jolly, R. A., Ciurlionis, R., Heindel, M., et al. (2002) Identifying toxic mechanisms using DNA microarrays: evidence that an experimental inhibitor of cell adhesion molecule expression signals through the aryl hydrocarbon nuclear receptor. *Toxicology* **181–182**, 537–550.
7. Wolfgang, G. H., Robertson, D. G., Welty, D. F., and Metz, A. L. (1995) Hepatic and adrenal toxicity of a novel lipid regulator in beagle dogs. *Fundam. Appl. Toxicol.* **26**, 272–281.
8. Grasso, P., and Hinton, R. H. (1991) Evidence for and possible mechanisms of non-genotoxic carcinogenesis in rodent liver. *Mutat. Res.* **248**, 271–290.
9. Hamadeh, H. K., Bushel, P. R., Jayadev, S., Martin, K., DiSorbo, O., Sieber, S., et al. (2002) Gene expression analysis reveals chemical-specific profiles. *Toxicol. Sci.* **67**, 219–231.
10. Kemper, B. (1998) Regulation of cytochrome P450 gene transcription by phenobarbital. *Prog. Nucleic Acid Res. Mol. Biol.* **61**, 23–64.
11. Voss, K. A., Liu, J., Anderson, S. P., Dunn, C., Miller, J. D., Owen, J. R., et al. (2006) Toxic effects of fumonisin in mouse liver are independent of the peroxisome proliferator-activated receptor alpha. *Toxicol. Sci.* **89**, 108–119.

12. Fielden, M. R., Brennan, R., and Gollub, J. (2007) A gene expression biomarker provides early prediction and mechanistic assessment of hepatic tumor induction by non-genotoxic chemicals. *Toxicol. Sci.* **99**, 90–100.

13. Nie, A. Y., McMillian, M., Parker, J. B., Leone, A., Bryant, S., Yieh, L., et al. (2006) Predictive toxicogenomics approaches reveal underlying molecular mechanisms of nongenotoxic carcinogenicity. *Mol. Carcinog.* **45**, 914–933.

14. Wallace, K. B., Hausner, E., Herman, E., Holt, G. D., MacGregor, J. T., Metz, A. L., et al. (2004) Serum troponins as biomarkers of drug-induced cardiac toxicity. *Toxicol. Pathol.* **32**, 106–121.

15. Singal, P. K., and Iliskovic, N. (1998) Doxorubicin-induced cardiomyopathy. *N. Engl. J. Med.* **339**, 900–905.

16. Kalivendi, S. V., Konorev, E. A., Cunningham, S., Vanamala, S. K., Kaji, E. H., Joseph, J., and Kalyanaraman, B. (2005) Doxorubicin activates nuclear factor of activated T-lymphocytes and Fas ligand transcription: role of mitochondrial reactive oxygen species and calcium. *Biochem. J.* **389**, 527–539.

17. Minotti, G., Menna, P., Salvatorelli, E., Cairo, G., and Gianni, L. (2004) Anthracyclines: molecular advances and pharmacologic developments in antitumor activity and cardiotoxicity. *Pharmacol. Rev.* **56**, 185–229.

18. Robert, J. (2007) Preclinical assessment of anthracycline cardiotoxicity in laboratory animals: Predictiveness and pitfalls. *Cell. Biol. Toxicol.* **23**, 27–37.

19. Chaiswing, L., Cole, M. P., St Clair, D. K., Ittarat, W., Szweda, L. I., and Oberley, T. D. (2004) Oxidative damage precedes nitrative damage in adriamycin-induced cardiac mitochondrial injury. *Toxicol. Pathol.* **32**, 536–547.

20. Jang, Y. M., Kendaiah, S., Drew, B., Phillips, T., Selman, C., Julian, D., and Leeuwenburgh, C. (2004) Doxorubicin treatment in vivo activates caspase-12 mediated cardiac apoptosis in both male and female rats. *FEBS Lett.* **577**, 483–490.

21. Gu, Y. G., Weitzberg, M., Clark, R. F., Xu, X., Li, Q., Lubbers, N. L., et al. (2007) N-{3-[2-(4-alkoxyphenoxy)thiazol-5-yl]-1-methylprop-2-ynyl}carboxy derivatives as acetyl-coA carboxylase inhibitors—improvement of cardiovascular and neurological liabilities via structural modifications. *J. Med. Chem.* **50**, 1078–1082.

22. Ganter, B., Tugendreich, S., Pearson, C. I., Ayanoglu, E., Baumhueter, S., Bostian, K. A., et al. (2005) Development of a large-scale chemogenomics database to improve drug candidate selection and to understand mechanisms of chemical toxicity and action. *J. Biotechnol.* **119**, 219–244.

23. Diel, P., Smolnikar, K., Schulz, T., Laudenbach-Leschowski, U., Michna, H., and Vollmer, G. (2001) Phytoestrogens and carcinogenesis-differential effects of genistein in experimental models of normal and malignant rat endometrium. *Hum. Reprod.* **16**, 997–1006.

24. Kuiper, G. G., Enmark, E., Pelto-Huikko, M., Nilsson, S., and Gustafsson, J. A. (1996) Cloning of a novel receptor expressed in rat prostate and ovary. *Proc. Natl. Acad. Sci. U. S. A.* **93**, 5925–5930.

25. Okada, A., Ohta, Y., Buchanan, D. L., Sato, T., Inoue, S., Hiroi, H., et al. (2002) Changes in ontogenetic expression of estrogen receptor alpha and not of estrogen receptor beta in the female rat reproductive tract. *J. Mol. Endocrinol.* **28**, 87–97.

26. Matthews, J., and Gustafsson, J. A. (2003) Estrogen signaling: a subtle balance between ER alpha and ER beta. *Mol. Interv.* **3**, 281–292.

27. Moriarty, K., Kim, K. H., and Bender, J. R. (2006) Minireview: estrogen receptor-mediated rapid signaling. *Endocrinology* **147**, 5557–5563.

28. Hall, J. M., and McDonnell, D. P. (2005) Coregulators in nuclear estrogen receptor action: from concept to therapeutic targeting. *Mol. Interv.* **5**, 343–357.

29. Filardo, E. J., and Thomas, P. (2005) GPR30: a seven-transmembrane-spanning estrogen receptor that triggers EGF release. *Trends Endocrinol. Metab.* **16**, 362–367.

30. Bjornstrom, L., and Sjoberg, M. (2005) Mechanisms of estrogen receptor signaling: convergence of genomic and nongenomic actions on target genes. *Mol. Endocrinol.* **19**, 833–842.

31. Levin, E. R. (2005) Integration of the extranuclear and nuclear actions of estrogen. *Mol. Endocrinol.* **19**, 1951–1959.

32. Houston, K. D., Copland, J. A., Broaddus, R. R., Gottardis, M. M., Fischer, S. M., and Walker, C. L. (2003) Inhibition of proliferation and estrogen receptor signaling by peroxisome proliferator-activated receptor gamma ligands in uterine leiomyoma. *Cancer Res.* **63**, 1221–1227.

33. Rhen, T., Grissom, S., Afshari, C., and Cidlowski, J. A. (2003) Dexamethasone blocks the rapid biological effects of 17beta-estradiol in the rat uterus without antagonizing its global genomic actions. *FASEB J.* **17**, 1849–1870.

34. Kazi, A. A., Jones, J. M., and Koos, R. D. (2005) Chromatin immunoprecipitation analysis of gene expression in the rat uterus in vivo: estrogen-induced recruitment of both estrogen receptor alpha and hypoxia-inducible factor 1 to the vascular endothelial growth factor promoter. *Mol. Endocrinol.* **19**, 2006–2019.

35. Li, X. H., and Ong, D. E. (2003) Cellular retinoic acid-binding protein II gene expression is directly induced by estrogen, but not retinoic acid, in rat uterus. *J. Biol. Chem.* **278**, 35819–35825.

36. Wu, X., Pang, S. T., Sahlin, L., Blanck, A., Norstedt, G., and Flores-Morales, A. (2003) Gene expression profiling of the effects of castration and estrogen treatment in the rat uterus. *Biol. Reprod.* **69**, 1308–1317.

37. Moggs, J. G., Tinwell, H., Spurway, T., Chang, H. S., Pate, I., Lim, F. L., et al. (2004) Phenotypic anchoring of gene expression changes during estrogen-induced uterine growth. *Environ. Health Perspect.* **112**, 1589–1606.

38. Naciff, J. M., Overmann, G. J., Torontali, S. M., Carr, G. J., Khambatta, Z. S., Tiesman, J. P., et al. (2007) Uterine temporal response to acute exposure to 17{alpha}-ethinyl estradiol in the immature rat. *Toxicol. Sci.* **97**, 467–490.

39. Heryanto, B., Lipson, K. E., and Rogers, P. A. (2003) Effect of angiogenesis inhibitors on oestrogen-mediated endometrial endothelial cell proliferation in the ovariectomized mouse. *Reproduction* **125**, 337–346.

40. Naciff, J. M., Overmann, G. J., Torontali, S. M., Carr, G. J., Tiesman, J. P., Richardson, B. D., and Daston, G. P. (2003) Gene expression profile induced by 17 alpha-ethynyl estradiol in the prepubertal female reproductive system of the rat. *Toxicol. Sci.* **72**, 314–330.

41. Naciff, J. M., Overmann, G. J., Torontali, S. M., Carr, G. J., Tiesman, J. P., and Daston, G. P. (2004) Impact of the phytoestrogen content of laboratory animal feed on the gene expression profile of the reproductive system in the immature female rat. *Environ. Health Perspect.* **112**, 1519–1526.
42. Helvering, L. M., Adrian, M. D., Geiser, A. G., Estrem, S. T., Wei, T., Huang, S., et al. U. (2005) Differential effects of estrogen and raloxifene on messenger RNA and matrix metalloproteinase 2 activity in the rat uterus. *Biol. Reprod.* **72**, 830–841.
43. Creasy, D. M. (1997) Evaluation of testicular toxicity in safety evaluation studies: the appropriate use of spermatogenic staging. *Toxicol. Pathol.* **25**, 119–131.
44. Stewart, J., and Turner, K. J. (2005) Inhibin B as a potential biomarker of testicular toxicity. *Cancer Biomarkers* **1**, 75–91.
45. Murugesan, P., Balaganesh, M., Balasubramanian, K., and Arunakaran, J. (2007) Effects of polychlorinated biphenyl (Aroclor 1254) on steroidogenesis and antioxidant system in cultured adult rat Leydig cells. *J. Endocrinol.* **192**, 325–338.
46. Adachi, T., Koh, K. B., Tainaka, H., Matsuno, Y., Ono, Y., Sakurai, K., Fet al. (2004) Toxicogenomic difference between diethylstilbestrol and 17beta-estradiol in mouse testicular gene expression by neonatal exposure. *Mol. Reprod. Dev.* **67**, 19–25.
47. Adachi, T., Ono, Y., Koh, K. B., Takashima, K., Tainaka, H., Matsuno, Y., et al. (2004) Long-term alteration of gene expression without morphological change in testis after neonatal exposure to genistein in mice: toxicogenomic analysis using cDNA microarray. *Food Chem. Toxicol.* **42**, 445–452.
48. Moustafa, G. G., Ibrahim, Z. S., Hashimoto, Y., Alkelch, A. M., Sakamoto, K. Q., Ishizuka, M., and Fujita, S. (2007) Testicular toxicity of profenofos in matured male rats. *Arch Toxicol.* **81**, 875–881.
49. Richburg, J. H., Johnson, K. J., Schoenfeld, H. A., Meistrich, M. L., and Dix, D. J. (2002) Defining the cellular and molecular mechanisms of toxicant action in the testis. *Toxicol. Lett.* **135**, 167–183.
50. Shultz, V. D., Phillips, S., Sar, M., Foster, P. M., and Gaido, K. W. (2001) Altered gene profiles in fetal rat testes after in utero exposure to di(n-butyl) phthalate. *Toxicol. Sci.* **64**, 233–242.
51. Tully, D. B., Bao, W., Goetz, A. K., Blystone, C. R., Ren, H., Schmid, J. E., et al. (2006) Gene expression profiling in liver and testis of rats to characterize the toxicity of triazole fungicides. *Toxicol. Appl. Pharmacol.* **215**, 260–273.
52. Rockett, J. C., Christopher Luft, J., Brian Garges, J., Krawetz, S. A., Hughes, M. R., Hee Kirn, K., et al. (2001) Development of a 950-gene DNA array for examining gene expression patterns in mouse testis. *Genome Biol.* **2**, 1–10.
53. Ryu, J. Y., Lee, B. M., Kacew, S., and Kim, H. S. (2007) Identification of differentially expressed genes in the testis of Sprague-Dawley rats treated with di(n-butyl) phthalate. *Toxicology* **234**, 103–112.
54. Page, B. D., and Lacroix, G. M. (1995) The occurrence of phthalate ester and di-2-ethylhexyl adipate plasticizers in Canadian packaging and food sampled in 1985–1989: a survey. *Food Addit. Contam.* **12**, 129–151.

55. Saillenfait, A. M., Payan, J. P., Fabry, J. P., Beydon, D., Langonne, I., Gallissot, F., and Sabate, J. P. (1998) Assessment of the developmental toxicity, metabolism, and placental transfer of di-n-butyl phthalate administered to pregnant rats. *Toxicol. Sci.* **45**, 212–224.

56. Mylchreest, E., Cattley, R. C., and Foster, P. M. (1998) Male reproductive tract malformations in rats following gestational and lactational exposure to Di(n-butyl) phthalate: an antiandrogenic mechanism? *Toxicol. Sci.* **43**, 47–60.

57. Thompson, C. J., Ross, S. M., and Gaido, K. W. (2004) Di(n-butyl) phthalate impairs cholesterol transport and steroidogenesis in the fetal rat testis through a rapid and reversible mechanism. *Endocrinology* **145**, 1227–1237.

58. Fukushima, T., Yamamoto, T., Kikkawa, R., Hamada, Y., Komiyama, M., Mori, C., and Horii, I. (2005) Effects of male reproductive toxicants on gene expression in rat testes. *J. Toxicol. Sci.* **30**, 195–206.

59. Pelletier, G., Li, S., Luu-The, V., Tremblay, Y., Belanger, A., and Labrie, F. (2001) Immunoelectron microscopic localization of three key steroidogenic enzymes (cytochrome P450(scc), 3 beta-hydroxysteroid dehydrogenase and cytochrome P450(c17)) in rat adrenal cortex and gonads. *J. Endocrinol.* **171**, 373–383.

60. Dalla Valle, L., Vianello, S., Belvedere, P., and Colombo, L. (2002) Rat cytochrome P450c17 gene transcription is initiated at different start sites in extraglandular and glandular tissues. *J. Steroid Biochem. Mol. Biol.* **82**, 377–384.

61. Kaplowitz, N. (2005) Idiosyncratic drug hepatotoxicity. *Nat. Rev. Drug Discov.* **4**, 489–499.

62. Liguori, M. J., and Waring, J. F. (2006) Investigations toward enhanced understanding of hepatic idiosyncratic drug reactions. *Expert Opin. Drug Metab. Toxicol.* **2**, 835–846.

63. Uetrecht, J. (2007) Idiosyncratic drug reactions: current understanding. *Annu. Rev. Pharmacol. Toxicol.* **47**, 513–539.

64. Waring, J. F., and Anderson, M. G. (2005) Idiosyncratic toxicity: mechanistic insights gained from analysis of prior compounds. *Curr. Opin. Drug Discov. Dev.* **8**, 59–65.

65. Fielden, M. R., and Halbert, D. N. (2007) Iconix Biosciences, Inc. *Pharmacogenomics* **8**, 401–405.

66. Leone, A. M., Kao, L. M., McMillian, M. K., Nie, A. Y., Parker, J. B., Kelley, M. F., et al. (2007) Evaluation of felbamate and other antiepileptic drug toxicity potential based on hepatic protein covalent binding and gene expression. *Chem. Res. Toxicol.* **20**, 600–608.

67. Dieckhaus, C. M., Thompson, C. D., Roller, S. G., and Macdonald, T. L. (2002) Mechanisms of idiosyncratic drug reactions: the case of felbamate. *Chem. Biol. Interact.* **142**, 99–117.

68. Naito, Y., Kuroda, M., Uchiyama, K., Mizushima, K., Akagiri, S., Takagi, T., et al. (2006) Inflammatory response of esophageal epithelium in combined-type esophagitis in rats: a transcriptome analysis. *Int. J. Mol. Med.* **18**, 821–828.

3

Quality Control of Microarray Assays for Toxicogenomic and *In Vitro* Diagnostic Applications

Karol L. Thompson and Joseph Hackett

Summary

The generation of high-quality microarray data for toxicogenomics can be affected by the study design and methods used for sample acquisition, preparation, and processing. Bias can be introduced during animal treatment, tissue handling, and sample preparation. Metrics and controls used in assessing RNA integrity and the quality of microarray sample generation are reviewed in this chapter. Regulations and guidelines involved in the application of microarrays as a commercial *in vitro* diagnostic device are also described.

Key Words: *in vitro* diagnostics; metrics; microarrays; quality control; standards.

1. Introduction

The application of microarray technology to toxicology began in earnest in the early 2000s with the publication of proof of concept studies that investigated whether genomic signatures could be used to correctly classify and predict chemical toxicities *(1,2)*. Genomic technology has since gained wider use in the early phases of drug development as a screen for the potential to induce organ-specific toxicities and in characterizing pharmaceutical mode of action. It is anticipated that there will be increased incorporation of toxicogenomics as supporting data for Investigational New Drug applications and increased use of microarray methodology to classify or monitor patients in clinical trials and in environmental risk assessments. For the effective implementation of this technology in clinical and regulatory settings, it is essential that suitable standards and metrics be established to effectively assess the performance of

From: *Methods in Molecular Biology, vol. 460: Essential Concepts in Toxicogenomics*
Edited by: D. L. Mendrick and W. B. Mattes © Humana Press, Totowa, NJ

microarray assays and ensure high quality of microarray data. In addition, the clinical use of microarrays is subject to governmental regulations for *in vitro* diagnostic devices.

Early studies that investigated the use of genomic technology to measure or predict toxicity did not focus on standards or controls but highlighted some areas of concern. The Institutional Life Sciences Institute (ILSI) Health and Environmental Sciences Institute (HESI) Technical Committee on the Application of Genomics to Mechanism-Based Risk Assessment conducted a series of multisite cross-platform studies to explore the feasibility of using genomics to identify toxicant-specific signatures for hepatotoxic, nephrotoxic, and genotoxic compounds *(3)*. Although common sets of gene changes reflective of site and mechanism of injury were observed across different sites, platforms, and independent studies, significant differences could be observed between sites using the same platform. Since these studies were conducted, there has been an evolution in microarray design, increased and improved annotation of the rat genome, and improvements in method optimization. Data comparability and reproducibility has been enhanced by the increased availability and use of qualified reagent kits and of automated systems for sample and array processing. A recent large consortial effort to compare microarray data results between platforms and sites showed high overall reproducibility can be achieved when probes are carefully mapped to curated cDNA sequence databases and data is generated in high-performing laboratories using standardized protocols *(4)*. However, significant differences between some array formats were observed. Commercial oligonucleotide arrays were shown to have superior performance to oligonucleotide spotted arrays prepared in core facilities in this and in previous studies *(4–6)*.

The role of quality control (QC) in microarray data generation is to ensure that results are reproducible and accurate. Standard methods for measuring assay and laboratory performance are not easily translatable to "omics" technologies. It is not feasible to construct a standard that can assess the specificity and sensitivity of all of the large number of end points that can be measured in a single "omics" assay. An additional challenge is present by the diversity in platforms, instrumentation, reagents, and protocols that are used to generate toxicogenomic (TG) data. Several guidelines for generating quality microarray data have been published that begin to address the need for establishing best practices for conducting microarray assays *(7–10)*. The need for standards and reference materials that can be used across different microarray formats is starting to be addressed by the microarray community under the collaborative efforts organized by the National Institute of Standards and Technology (NIST) and the Food and Drug Administration (FDA) *(11,12)*. As public repositories of toxicogenomic data become more highly populated, there is an increased need for methods and standards to ensure data quality so that these data sets can be reliably used to create an open access toxicogenomics knowledge base *(13)*. In addition to standards, performance monitoring programs can be implemented to address the need for demonstrating the competency of microarray core facilities and contract research organizations to clients and for Clinical Laboratory Improvement Amendments (CLIA) certification of clinical laboratories *(14)*.

Quality control measures can be applied to each of the steps involved in the generation of TG samples. These steps are study design, animal and tissue handling, RNA isolation, microarray sample processing, sample hybridization, microarray scanning, and data analysis (**Fig. 1**). The performance of some of these steps can be monitored with either specific control reagents and/or metrics. **Sections 2 to 6** in this chapter outline processes within the first six of these steps that have been identified to contribute to data variability and summarize approaches that have been used to ensure sample and data quality. Microarray data analysis is addressed in a separate chapter in this volume. Although this chapter is focused on QC measures that apply to samples generated from rats,

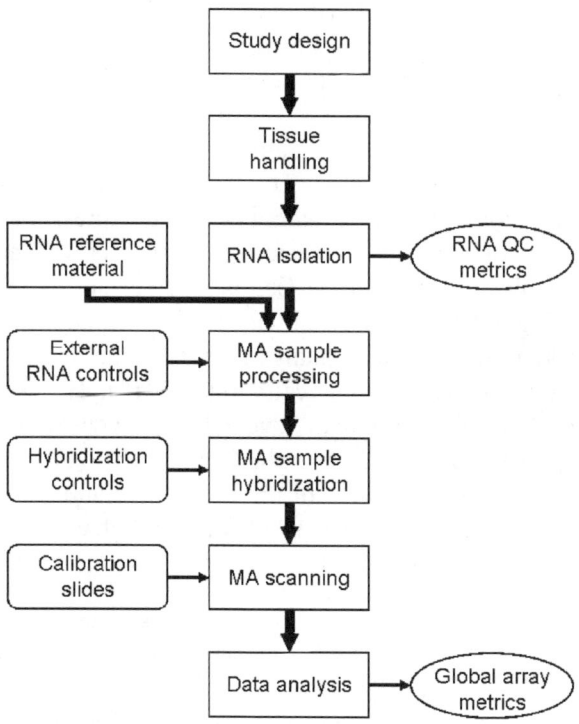

Fig. 1. Outline of the steps involved in a toxicogenomic experiment where QC metrics and controls can be applied.

the dominant preclinical model system, many are also applicable to samples from other mammalian species, including clinical samples.

One application of toxicogenomics is the identification of biomarkers that can potentially bridge from preclinical to clinical application. Currently, there is a limited number of approved pharmacogenomic *in vitro* diagnostic (IVD) devices, although pharmacogenomic information is currently included on 10% of approved drug labels *(15)*. **Section 7** in this chapter outlines the regulatory requirements involved in the application of microarrays as diagnostic devices.

2. QC of Sample Acquisition and Preparation for Toxicogenomic Studies

2.1. Sample Acquisition

In an optimally designed experiment, control and treated groups are matched in all aspects of animal care and manipulation except for the experimental variables under investigation. Assessment of toxicant-specific effects on gene expression can be confounded if there are significant differences between biological or technical replicates within a study that are due to nonuniform handling of animals or samples. Experiments should be designed to avoid the introduction of bias in the data due to nonrandomized processing of treatment groups.

Large-scale TG databases are typically designed using high-standardized study protocols *(16)*. However, most TG studies are variable in design, and the impact of variations in study protocol on gene expression in control and treated animals has not been well characterized. A project was recently initiated by the HESI Technical Committee on Genomics to establish a publicly accessible database of annotated control animal microarray data that will provide a resource for identification and analysis of baseline fluctuations in gene expression due to biological or technical variables *(17)*. The possible effect of experimental protocol variables like diet, vehicle type, vehicle route of administration, dosing frequency, method of sacrifice, and the use of anesthetics during the study need to be examined. One of the goals of establishing a database of baseline gene expression in the rat is to identify and characterize the inherent variability due to differences in study design prior to introduction of drugs or toxicants.

2.1.1. Bias Introduction Through Biological Variance

A microarray experiment creates a snapshot of the dynamic process of gene expression within an organism. Laboratory mice and rats, like other mammals, develop a circadian rhythm in response to a fixed light-dark photoperiod that entrains circadian pacemakers in the hypothalamic suprachiasmatic nucleus (SCN). Genes expressed in peripheral tissues have been identified that also undergo circadian cycles in abundance *(18)*. It has been estimated that 10% of genes expressed in liver and 8% of genes expressed in heart show circadian regulation in mice *(19)*. However, there are significant differences in circadian time of peak expression and limited overlap between the sets of genes exhibiting circadian regulation between these two tissues. Of relevance to toxicogenomics, transcript levels of certain genes involved in metabolism and cholesterol synthesis exhibit a diurnal pattern of expression in liver and kidney in rats acclimated to a light/dark cycle *(20,21)*. Marked changes in expression of a subset of genes in rat liver were observed between time points separated by only 6 h.

Fasting and circadian rhythm have both unique and interacting effects on gene expression in the rat. Rats are often fasted during toxicity studies to increase sensitivity to hepatotoxicity because fasting decreases glutathione levels and metabolic enzyme induction *(22)*. In a recent analysis of control animals that differed both in respect to overnight fasting and in time of day of sacrifice (by 6 h), liver genes involved in lipid metabolism, energetics, and circadian rhythm were primarily affected *(23)*. The identified gene changes were consistent with decreased glycogen breakdown and glycolysis and increased lipid metabolism, reflective of the use of fatty acids as an energy source during fasting. In an independent study, subsets of liver genes were identified that exhibited a significant difference in day/night ratio only in nonfasted animals, only in fasted animals, or irrespective of feeding status *(20)*. For example, genes involved in glucose use were diurnally regulated in the liver but mostly unaffected by food restriction. Fasting had less of an effect on diurnally regulated genes in kidney, which tended to be distinct from genes identified as diurnally expressed genes in liver.

These findings on the effect of the circadian rhythm on gene expression underline the importance of necropsy time of day in TG studies. Time-matched controls are recommended for all study designs, including exposures less than 24 h to reduce the introduction of bias due to differences in diurnally expressed genes. In addition, if a large study is terminated on one day, the order of animals to be necropsied should be randomized among treatment groups to minimize bias due to changes in diurnally expressed genes. Additionally, toxicant treatments that affect diet consumption patterns may indirectly alter liver gene expression changes. Limiting access to food has been shown to entrain the circadian clock gene *Per1* in peripheral organs (liver and lung), independent of the SCN *(24)*.

Several recent studies have examined the effect of vehicle and anesthesia on gene expression in control animals *(25,26)*. Relative to methylcellulose, the use of corn oil as a vehicle was found to modulate genes involved in cholesterol or fatty acid metabolism in liver. For a number of genes, an inverse effect on expression was observed at shorter time points (3,6,9, or 24 h) after a single dose compared with the effect after repeated daily doses (for 4, 8, 15, or 29 days). The observed changes were reflected in blood biochemistry measurements and were based in part on the effect of corn oil on food consumption *(25)*. Exposure to sevoflurane for 2 h was found to affect a small number of transcripts in the six tissues examined relative to 0-h controls and included a number of known circadian genes *(26)*. The greatest effect observed in this study was on liver gene expression after 6 h inhalation exposure to sevoflurane, which may not be relevant to most TG study protocols.

Individual animal variability in gene expression is low among control groups of laboratory rats under controlled environmental conditions *(27)*. However, the level of individual variability among non-rodent laboratory animals like the dog and monkey should be determined prior to experimental design so that appropriate group sizes are used or that limitations in statistical power are understood. The confounding effects of variation between individual subjects was demonstrated in a study of gene expression in human blood, where different expression patterns were associated with gender, age, time of day at which sample was taken, and with inherent individual variability (i.e., genotype, epigenetic, or environmental factors) *(28)*.

2.1.2. Bias Introduction Through Differences in Tissue Handling

TG studies usually form one of several data streams that are integrated together for the interpretation or prediction of toxicologic outcomes *(29)*. The tissues under investigation in a TG study can be source material for genomic, histopathologic, pharmacokinetic, and proteomic measurements. Therefore, an entire intact organ is often not available for RNA extraction, and bias could be introduced by sampling different regions of tissue between or among control and test animals. This concern also applies when sampling is limited to a region thought to be most sensitive to injury in a particular organ. Although liver is considered to be one of the more homogenous tissues, differences in gene expression in response to acetaminophen have been observed between the left and median lobes of F344 rat liver *(30)*. In a study of furan-mediated hepatotoxicity, an increased severity of lesions and a greater number of gene expression changes were observed in the right and caudate lobes of the liver compared with the left and medial lobes *(31)*. These results suggest that histopathology and gene expression results could be incongruent if both analyses are not performed on the same liver lobe.

Tissues are commonly preserved for subsequent RNA isolation either through immediate freezing in liquid nitrogen or by immersion in tissue collection and stabilization solutions, for example, Ambion RNA*later*® (Austin, TX), which allows for preservation of tissues at ambient temperatures. Tissues are particularly sensitive to RNA degradation during the transition between storage at –70°C and homogenization in chaotropic buffer *(32)*, a step that is avoided with the use of tissue stabilization solutions. The impact of RNA degradation on microarray sample quality is discussed in **Section 2.2**.

The use of archival tissue for gene expression analyses is a resource of great potential value for retrospective analysis of the mechanism of drug-induced toxicities. However, tissue fixative and processing methods have a negative impact on RNA integrity (reviewed in Ref. *33*). Recommended methods for the extraction and use of RNA from formalin-fixed, paraffin-embedded tissue have been reviewed by Lewis et al. *(34)*.

It is sometimes desirable to isolate RNA from specific regions using laser capture microscopy (LCM) or other procedures. On account of the small sample size, microdissected samples usually need to undergo additional rounds of amplification during target synthesis, which tends to create products that are truncated at the 5' end. The effect of the LCM procedure on detection of gene expression change induced by clofibrate was examined by Michel et al. *(35)*. LCM processing tended to mute the lower-fold changes that were significantly induced by clofibrate treatment in whole liver, although the molecular signature associated with peroxisome proliferator exposure was still observable. Microdissection can be successfully used to isolate samples for microarray analysis, but some impairment of sample quality should be expected and comparisons may be limited to control samples prepared using the same techniques. Additionally, this method can introduce a higher degree of technical variability than what is usually observed between technical replicates *(36)*. The loss of sensitivity based on technical limitations from using samples isolated using LCM should be weighed against the increased sensitivity of analyzing only the tissue region associated with toxicity.

Gene expression profiling of peripheral blood is of interest for TG studies because of potential translation to the clinical setting. When peripheral blood is the tissue source of RNA for genomics, the primary concerns are that (1) the sample is adequately preserved for RNA extraction and (2) blood components known to interfere with microarray results are effectively removed. Unlike tissues, blood is usually processed for RNA without freezing. Prolonged storage of blood before processing has been shown to reduce sample comparability. Total blood may have decreased sensitivity compared with blood fractionated for peripheral blood mononuclear cells (PBMCs), which are the most transcriptionally active cell population in blood, but the fractionation process may itself introduce bias into the process. It has been well documented that significant and rapid changes in cytokine mRNA abundance occur during *ex vivo* incubation of peripheral blood *(28,37–39)*. In total RNA isolated from whole blood, there is a dominating contribution of globin mRNA from the reticulocyte population due to the abundance of red blood cells (RBCs) in whole blood.

Some common techniques for preparing blood for gene expression analysis include isolation of RNA from whole blood (e.g., PAXgene™), selective lysis of erythrocytes prior to RNA isolation (e.g., QIAamp), and separation of PBMCs prior to RNA isolation (e.g., Ficoll-Hypaque method, BD Vacutainer-CPT sodium citrate tubes). Methods like PAXgene are often used because they stabilize samples for several days prior to RNA isolation and dampen the spontaneous cytokine inductions seen after storage of whole blood in EDTA tubes *(39)*. Hemolysis, which is induced during some methods of RNA isolation from blood, may induce changes in genes such as interferon gamma. Additionally, overnight storage of blood has a significant effect on gene expression compared with samples drawn and processed immediately *(37)*.

A comparison of the effects of different blood isolation techniques on the quality of results produced on Affymetrix GeneChip® arrays (Santa Clara, CA) has been conducted *(38)*. Differences in expression profiles seen were attributed to the preferential isolation of different blood cell populations by each method. The blood cell population in a sample can be determined by the expression level of signature genes for RBCs, granulocytes, and PMBCs. These reference genes could be used to ensure that samples contain comparable levels of blood cell components prior to investigation of treatment- or disease-induced changes. Techniques that do not remove globin mRNA tended to have reduced sensitivity and increased variability compared with the other methods examined. The use of a protocol that reduces globin mRNA in whole human blood samples can improve the quality of results on Affymetrix arrays with total RNA from whole blood isolated with the PAXgene method *(40)*. In summary, these observations indicate blood collection, storage, and RNA isolation need to be performed under standardized conditions if data is to be integrated across laboratories.

It has been shown in several studies that *ex vivo* incubation of tissues or blood samples prior to RNA extraction (e.g., from delay in sample processing) induces a distinct set of genes linked to early response, hypoxia, and other stress responses *(37,38,41)*. Downregulation of genes involved in other cellular functions (apoptosis, metabolism, and immune response) can also be observed. In summary, investigators need to be aware of the effect of processing method, protocol, and environment on the variability of genomic data.

2.2. RNA Quality Metrics

High-quality RNA is important for achieving reproducible and interpretable results on microarrays. RNA quality may be best assessed by an RNA quality index that takes into account several different physical metrics *(42)*. RNA purity and integrity are important qualities for generating the best results on microarrays. The ratio of optical density at 260 nm and 280 nm is the standard measurement of RNA purity. Pure RNA should have an A_{260}/A_{280} ratio of 2.0. The lowest acceptable ratio for microarray analysis is, however, a judgment call based on laboratory experience. Contamination of samples with protein, genomic DNA, or chemicals can interfere with accurate quantitation or with downstream applications. DNA contamination is not easy to detect, so some protocols for RNA sample preparation for microarray experiments include treatment with deoxyribonuclease I prior to reverse transcription.

RNA quantitation is important to ensure the consistent use of an amount of starting material that is appropriate for a given labeling method. Although the relationship between one A_{260} unit and 40 µg/ml RNA specifically applies to measurements in water, the A_{260}/A_{280} ratio is significantly affected by pH and should be performed in a buffered solution. Spectrophotometers that can quantitate samples of 1 µl volume may help reduce the variability and inaccuracy in RNA measurements by circumventing the need for sample dilution that is involved with the use of cuvettes.

Microfluidics-based platforms for nucleic acid analysis, such as the Agilent 2100 Bioanalyzer (Wilmington, DE), have become standard technologies for assessing RNA integrity in microarray experimental protocols. This technology, which involves automated electrophoretic separation of RNA and quantitation by fluorescence detection, provides a calculation of the 28s/18s rRNA ratio and an electrophoretic trace of the RNA sample.

Intact total RNA has sharp peaks of 18s and 28s rRNA. The ratio of these rRNA peaks is often used as a measurement of RNA integrity. However, 28s/18s rRNA ratios lower than 2.0 are often found in total RNA samples in the absence of any visible degradation products *(43)*. In one study of RNA metrics correlative of high-quality microarray data, 28s/18s rRNA ratios were not found to be predictive of sample quality *(8)*. This metric is also not a useful indicator of sample integrity if total RNA is partially purified prior to target preparation *(16)*. Recently, a software algorithm has been developed for use in combination with microfluidics-based platforms to provide more consistency in assessment of RNA integrity *(44)*. The software program assigns an RNA Integrity Number (RIN) to each sample based on the characteristics of multiple informative segments of an electropherogram. Similar calculations can be made with open source software *(45,46)*.

There is increasing interest in requiring inclusion of electropherograms in data submitted to public databases and in other instances of microarray data exchange *(43)*. As a metric, an electropherogram provides a more complete picture of RNA quality than the 28s/18s rRNA ratio alone *(43)*, and the RIN provides a quantitative summary metric of this reading. RNA with RIN <6 was shown to be of poor quality for quantitative reverse transcription–polymerase chain reaction (qRT-PCR) experiments *(44)*, but the lower limit in acceptable RIN value for RNA that will be used in microarray studies may depend on the purpose of the study. A lower level of RNA quality will tend to reduce the statistical power of a study because of increased measurement error. The signal levels of some transcripts are much more sensitive to RNA than others, attributable in part to probe distance from the polyadenylation (polyA) site at the end of the corresponding target sequence *(32)*.

The expected yield of RNA from a given weight of tissue is an additional quality check of the effectiveness of an RNA isolation protocol. These values can be found in literature supplied by companies that provide RNA isolation reagents (see, e.g., Ref. *47*). For tissues that are fibrous like heart and skeletal muscle, yield and quality may be affected if thorough disruption of the tissue and/or removal of abundant contractile proteins (e.g., by proteinase K digestion) are not performed. Tissues rich in nucleases and nucleic acids like spleen, thymus, and pancreas are highly sensitive to degradation. Additional precautions should be taken to ensure that ribonucleases are effectively inhibited and that sample viscosity is reduced. Tissues with high lipid content, like brain, may require preprocessing steps to facilitate lysis of fatty tissues. A lower than expected yield in RNA from a tissue should be considered a red flag for proceeding further with the sample.

Most labeling methods use the polyA tail of RNA as a template for priming cDNA synthesis. If there is RNA degradation in a sample, there will be fewer full-length cDNAs created. On platforms where probes are designed to be 3' biased, there is a reduced requirement for intact RNA. On Affymetrix GeneChip arrays, probes derived from the 5', middle, and 3' portions of certain universally expressed genes (e.g., beta-actin and glyceraldehyde-3-phosphate dehydrogenase [GAPDH]) are included in the array design to serve as indicators of sample degradation. The 3' to 5' ratios of these transcripts are included in the Expression Report file for each array generated by the MAS5.0 algorithm. A 3' to 5' GAPDH ratio of less than 3 is considered to be indicative of undegraded RNA *(7)*. However, these quality parameters cannot be applied when using protocols that involve processing samples captured by LCM and small sample amplification by multiple rounds of cDNA synthesis.

3. QC of Microarray Sample Processing

Data comparability is enhanced by the use of protocols and reagents optimized to achieve reproducible results and to adherence to protocol standardization across laboratories. In addition, data was found to be more reproducible on commercial microarrays compared with in-house spotted arrays *(4,5)*, probably due to more consistent manufacturing practices and application of more advanced levels of quality assessment. Technical proficiency in generating microarray data is assessed through the use of sample metrics that reflect the overall performance or that of individual steps in the process and through the use of internal or external controls whose measurements serve as surrogates for the performance of test samples.

3.1. Checkpoints for Process Monitoring

The first step in processing RNA for hybridization to arrays is cDNA synthesis. This step is usually followed by amplification with T7 RNA polymerase to produce cRNA. The efficiency of the cDNA and cRNA reactions can be monitored by the yield and the size of products generated. Most commonly, only metrics from the cRNA synthesis are measured and recorded if that is the reaction end point. Yield and size can be determined using spectrophotometers and microfluidics-based platforms that read samples of 1 µl in volume. The expected yield will be dependent on the reagents used and should be provided by the manufacturer. Typically, products from a good-quality cDNA synthesis range from 500 to 3000 bp *(7)*. On arrays, hybridization of cDNA or cRNA to probes that are complementary to sequence that is located far upstream of the 3' end of transcripts will be reduced if reverse transcription reactions are incomplete or the RNA template is degraded. In protocols that use direct labeling (i.e., through incorporation of nucleotides conjugated to fluorescent dyes into cDNA or cRNA), the reaction yield is also quantitated by measuring sample fluorescence. For protocols that use indirect labeling (i.e., incorporation of nucleotides conjugated with biotin, digoxigenin, or aminoallyl-modified), the amount of label incorporation is not usually measured. In many protocols, cRNA samples are fragmented prior to hybridization. The success of the fragmentation step can be monitored by a shift in the size of products to 50 to 200 nt using either microfluidics-based platforms or gel electrophoresis.

Some array formats are designed for use with two-color labeling protocols, which involve the incorporation of different fluorescently labeled nucleotides into treated and reference or control samples to allow simultaneous hybridization of both samples to the same array. The use of different dyes introduces a bias in the experiment due to different rates of dye incorporation into the sample and differences in quantum efficiency between the two dyes.

The use of indirect labeling methods, in which, for example, the fluorescent label is introduced into the sample through coupling to aminoallyl-modified nucleotides that are incorporated into cDNA or cRNA, only corrects for dye bias that results from differential dye incorporation. Dye bias can also result from differential sensitivity of Cy5 and Cy3 dyes to quenching, photobleaching, or degradation. The environmental ozone sensitivity of certain cyanine dyes (Cy5, Alexa 647) may track as a seasonal or diurnal effect on microarray data quality *(48)*. Some systematic dye bias effects can be controlled by running replicate arrays in which the orientation of dye incorporation is switched between treated and control samples.

3.2. External Controls for Process Monitoring

External controls are essential reagents for use in the qualitative assessment of different aspects of technical performance involved in generating microarray data (i.e., sample labeling, hybridization, and grid alignment). These controls are provided by many array manufacturers and consist of non-mammalian sequences that are spiked into samples and hybridize to corresponding control probe sequences on their arrays. Sets of spike-in targets and probes are also commercially available for use on in-house spotted arrays. External controls tend to be selected from prokaryotic or plant gene sequences that do not cross-hybridize to mammalian transcripts or probes and are usually not inter-changeable between platforms. A recent study evaluated the application of external RNA controls for four different commercial array formats *(49)*. Although the currently available spike-in controls are limited in concentration range, control RNAs currently in development by the External RNA Control Consortium (ERCC) are designed to be more extensive and applicable across platforms *(11)*. The ERCC controls are the primary resource that the proposed protocols were designed around that were recently issued by the Clinical and Laboratory Standards Institute for the use of external controls in microarray and qRT-PCR experiments *(50)*.

Labeling controls are polyA RNAs that are spiked into RNA samples before the reverse transcription step to monitor labeling and subsequent processes. These are usually provided as sets of polyadenylated prokaryotic RNAs and are used for qualitative assessments. Some platforms also contain probes for negative controls that can be used to calculate the noise threshold and to flag probes with signals below the lower limit of detection. Signal uniformity across arrays can be assessed if the array design includes multiple replicate spot locations for control sequences that are distributed across the array grid.

External controls that are added after target synthesis assess the success of the hybridization and staining steps. For Affymetrix arrays, hybridization controls are supplied as a kit of four labeled prokaryotic gene transcripts that span a concentration range. Optimally, all hybridization probe signals should be detected as present calls using the Affymetrix MAS5.0 signal detection algorithm. Additional controls include those that are used to identify the gridding of probes on microarrays and allow for proper grid alignment on images. These controls can be designed to bind to the boundaries of the probe area on an array (e.g., Oligo B2 on Affymetrix arrays) or to control probes colocalized with every feature on the microarray (e.g., LIZ-labeled internal control probes on Applied Biosystems arrays (Foster City, CA)).

4. QC of Microarray Scanning

Fluorescence standards for microarray scanners can be used to assess the performance limits of a scanner and allow discrimination between hybridization failures and scanner defects. Fluorescence calibration slides for scanners have dilution series of dyes (Cy3 and Cy5, or dyes with similar spectral properties) that allow for evaluation of the linear dynamic range of the scanner and of channel cross-talk as a function of dye concentration. These controls are currently available from only a few commercial sources. In addition, software programs that verify scanner performance are available for some scanners.

Typical array scanners have an output range between 0 and 65,535 relative fluorescence units for each pixel. The fixed dynamic range of fluorescence limits the extent of fold-change differences that can be observed between two samples *(51)*. This dampening effect is most apparent at the upper and lower limits of fluorescence detection, so it is important to note that optimal results may not be achieved in these areas.

Photomultiplier tube (PMT) setting was one of the major sources of variability among the Affymetrix array data generated in the multisite toxicogenomics project conducted by the HESI Genomics Committee *(52)*. Earlier models of scanners for Affymetrix arrays allowed for adjustment of PMT settings (high or low). Newer models have a fixed gain, so this source of variability is only applicable to data generated with the older models but remains a potential issue in comparisons of archival data with more recently generated data.

After scanning an array, the image should be inspected for artifacts such as scratches, bubbles, and high regional or overall background. Inspection can be visual or automated, using available software programs. Image analysis programs often have the ability to identify features with saturated pixels and other nonuniformities in a scanned image, and these outliers can be excluded from downstream analyses.

5. Global Quality Metrics for Microarray Data

After signal intensities are calculated from scanned images, one common method of assessing data quality is to compare the signal data from each array to that obtained from technical or biological replicates or against historic results. For example, for population of a large-scale toxicogenomics database, samples were excluded that had correlation coefficients below 0.8 in pairwise comparisons with matched biological replicates or with a tissue standard composed of pooled control samples for the same tissue *(16)*. Additionally, the similarity of samples within and between treatment groups can be visualized by applying principal component analysis to the signal data. Both of these methods are measurements of data precision, not accuracy.

Another commonly used quality index is comparison of the number of probes that are above a threshold value on a hybridized array with results typically obtained from similar RNA sources. On Affymetrix arrays, this index is captured in MAS5.0 Expression Report Files as Percent Present Calls and has been shown to be useful for monitoring performance over time by a core facility *(53)*. A present/absent detection call is made by a statistical algorithm within the Affymetrix MAS5.0 software using perfect match and mismatch probe pairs on the arrays *(54)*. It has been recommended that Present Calls should vary by less than 10% between samples in the same project *(7)*. As mentioned previously, some platforms use negative control probes to estimate global or local background on a microarray. Features with signals that are not significantly above background are flagged by the feature extraction software. The number of these outliers is reported as part of the summary statistics for an array and can be used as part of a strategy to exclude data of poor quality.

Array-level QC metrics can also be obtained using alternate algorithms for calculating signal values from Affymetrix arrays. dChip and the Bioconductor packages affy and affyPLM have algorithms that identify outliers among groups of arrays. In a recent study, the relationship between these and other "post-chip" metrics, and their correlation with four "pre-chip" metrics (postmortem interval, subjective and objective [RIN] RNA quality assessments from electropherogram traces, and cRNA yield) was evaluated for a data set composed of >100 human brain samples run on Affymetrix U133A GeneChips *(55)*.

6. RNA Microarray Standards for Performance Testing

In addition to the platform-specific external RNA controls described previously, there are consortial efforts under way to provide RNA reference materials that are designed to assess performance on microarrays across formats. Three aspects of assay performance that are important to establish are the limits of accuracy, precision, and linear range. To properly assess performance, a reference material should contain well-characterized levels of target transcripts with signals that span the dynamic range of measurement. This type of standard is important for measuring laboratory proficiency and process drift and for assessing the effect of changes in methods, reagents, and data analysis.

A set of external RNA controls is being developed by the ERCC that will consist of approximately 96 well-characterized polyadenylated transcripts composed of random unique and non-mammalian sequences *(11,56)*. These controls are designed to be added into a test sample or can be tested in a neutral background. Manufacturers of most commercial microarrays have agreed to issue new versions of their arrays that include probes for these control sequences. After validation, it is planned that these controls will be available in plasmid vectors from a nonprofit source like ATCC (American Type Culture Collection) and as quantified RNAs through a commercial source.

Scientists at the Center for Drug Evaluation and Research have led a collaborative effort with other government agencies and industry to design and test a reagent for use in proficiency testing, process drift monitoring, protocol optimization, and other performance assessments on rat whole-genome expression microarrays *(12)*. The reagent is composed of a set of two mixed-tissue RNA samples formulated to contain different known ratios of RNA for each of four rat tissues (brain, liver, kidney, and testes). The two mixes form the basis of an assay that measures the ratios of a set of reference probes that bind tissue-selective gene transcripts identified on Affymetrix, CodeLink™ (Applied Microarrays, Tempe, AZ), and Agilent arrays. The reference material is regenerable and can be prepared from pooled RNA from control rat tissues prepared in-house or from commercial sources. With this material, assay performance (sensitivity and specificity) can be measured using area under the curve (AUC) as a metric from receiver-operator characteristic (ROC) plots (Pine et al., in preparation).

7. Regulatory Requirements for Commercialization of *In Vitro* Diagnostic Devices

After gene expression experiments as previously described have been completed in humans, an association of certain genes (e.g., 10, 100, 1,000, etc.) with the toxic condition might be identified. This association would then need to be validated. If validation results support the association of the selected set of genes with the toxic condition in question, a decision could be made to commercially market the set as an IVD device. For example, an IVD might be marketed as a test to identify patients considered for treatment with a particular drug. Another example is that those testing "positive" with the selected gene set would be more likely to experience a toxic reaction if they received the drug. Those that test "negative," might be good candidates for the drug.

Within the FDA, certain regulatory and scientific requirements exist that must be met before the commercial distribution of an IVD (http://www.fda.gov/cdrh/devadvice/). By federal statute, IVDs are considered medical devices *(57)*. All medical devices are subject to general control regulations. General controls consist of labeling requirements for the device, registration of the firm, and listing of the device. Manufacturers of devices are also subject to good manufacturing practices (GMPs), as well as the provisions addressing adulteration and misbranding of the device, and being banned from distribution. In addition, the manufacturer must provide certain notifications and is subject to keeping records and providing reports. Additional information is available from the Division of Small Manufacturers, International, and Consumer Assistance (MSMICA) *(58)* (http://www.fda.gov/cdrh/industry/support/).

All medical devices are determined by the FDA to be in one of three regulatory categories or classes: I, II, or III. These classes are associated with the risk to the patient or subject being tested. For an IVD, the risk is related to the effect that the results of the test could have on patient management. Class I devices receive the least amount of FDA oversight and are subject to general controls. Class II devices receive more FDA oversight than the class I devices. For class II devices, general controls alone are not considered sufficient to provide reasonable assurances of safety and effectiveness of the device. Therefore, information about the IVD would exist to enable special controls to be established. Class III devices receive the most intense FDA oversight as insufficient information exists to determine that general controls will provide reasonable assurances that the device is safe and effective for its intended use, or the device is of substantial importance in preventing impairment to human health, or presents a potential unreasonable risk of illness of injury.

For example, an IVD used as a diagnostic test to detect cancer would be a high-risk device because a false-positive result could lead to exposure to radiation, therapy with an antineoplastic agent with serious side effects from the agent, or exploratory surgery. These are regarded as class III IVDs, requiring a PreMarket Application (PMA) approval before they can be commercially marketed. On the other hand, if the patient has already been diagnosed as having cancer, and has been treated, then a test to detect the reoccurrence of the cancer would be at a lower risk. This lower risk is based on the presumption that the test is only part of the comprehensive follow-up of a patient with cancer and that physicians are aware that analytical and biological false associations between reoccurrence and test results are well established.

If a decision is made to market an IVD, then depending on the class, one of two types of applications are made to the FDA. Class I and II devices are subject to a premarket notification, more commonly referred to as a 510(k) submission.

This name comes from section 510(k) of the FD&C act. Class III devices are subject to a PMA, which is similar to a New Drug Application (NDA). Most class I devices are exempt from the requirement to submit a 510(k). Premarket 510(k) submissions are required even for class I devices when these devices are for (1) use in diagnosis, monitoring, or screening of neoplastic diseases (IVDs using immunohistochemical procedures are exempt from the requirement to file a 510(k), however); (2) the screening or diagnosis of genetic disorders; (3) surrogate markers to screen, monitor, or diagnose life-threatening disease; (4) monitoring therapy; (5) assessing the risk of cardiovascular disease or for diabetes management; (6) identifying a microorganism directly from clinical material; (7) detection of antibodies other than IgG to microorganisms, unless the antibodies are for results that are not qualitative, are used to determine immunity, or used in matrices other than serum or plasma; and (8) invasive testing, or point-of-care testing *(59)*.

All class II devices are subject to the 510(k) requirements *(60)*, and all class III devices are subject to the PMA requirements *(61)*. If a device is not available in finished form and is not offered for commercial distribution, a 510(k) submission is not required, however.

7.1. Premarket Notification/510(k) Submission

Class II devices and class I devices that are not exempt by regulation from a 510(k) submission must have a FDA-cleared 510(k) to be commercially marketed. Among the information contained in the submissions are data demonstrating that the new device is substantially equivalent in performance to a legally marketed device for the same intended use. Both analytical and clinical sensitivity and specificity are determined and reported in the device's labeling. Information summaries on recently cleared 510(k) submissions are available at the Office of In Vitro Diagnostic Devices Evaluation and Safety (OIVD) Web page *(62)* and at the Center for Devices and Radiological Health (CDRH) Web page *(63)*.

7.2. Premarket Approval Submission

Class III devices must be shown to be safe and effective on their own merits. If a new device is compared with a legally marketed device for the same use, the data is only regarded as ancillary information. The studies that are performed to show a class III device to be safe and effective involve comparison of the accuracy of the device to clinical diagnosis as determined by the physician. As in the case of a 510(k) submission, both analytical and clinical sensitivity and specificity are determined and reported in the device's labeling. Summaries of the PMA studies for IVDs are also available on the OIVD Web page. Before the FDA approves a PMA, the manufacturing site must first pass a GMP inspection.

7.3. Investigational

When a new drug is being evaluated in humans, an Investigational New Drug (IND) application is required to be approved by the FDA before the drug can be administered for evaluation. A similar procedure exists for medical devices where Investigational Device Exemptions (IDEs) are filed. For IVDs, however, an IDE is rarely required because one of the reasons a diagnostic device would be exempt from the IDE requirements is when the new device is used for diagnostic procedures and the result is confirmed by another medically established diagnostic product or procedure *(64)*. The agency receives one to three IDEs for IVDs each year.

7.4. Early Communication (Pre-IDE)

The FDA does offer an informal nonbinding opportunity for communication between the agency and a manufacturer of a new IVD before the device is evaluated with clinical samples. This is the pre-IDE process *(65)*. As part of the pre-IDE process, a developer may meet informally and present to the FDA information on the IVD being developed. Based on the discussions at the meeting, the agency gains an earlier understanding of the device and the developer a better understanding of the regulatory process. Also based on the discussions, the developer would then submit a pre-IDE for feedback and advice from the agency. Within 60 days, the agency would provide a formal reply. The reply would identify what is acceptable to the FDA for gathering and analyzing the data from the investigational study. But this does not guarantee the study will produce useful results. Rarely does the pre-IDE process result in a formal IDE submission.

Whereas the agency charges a fee for review of 510(k) and PMA submissions, no fee is assessed for pre-IDE reviews. If the developer follows the agency's suggestions and recommendations, when the 510(k) or PMA is submitted a more efficient and timely review process should result. However, if the developer does not follow the FDA recommendations, this can result in delays, requests for clarifications, and frustration on both sides. Therefore, it is beneficial to both parties if the FDA recommendations are followed. The agency processes approximately 200 pre-IDEs per year for all IVDs.

Although the agency has not cleared or approved any toxicogenomic IVDs, it does have a long history of clearing IVDs that are used to monitor levels of various drugs such as gentamicin and digitoxin to ensure appropriate therapy and neuroleptic drugs to determine if a patient is at the appropriate level of the drug. IVDs have also been cleared for methadone assays in patients for treatment or overdose detection. Regulations identifying clinical toxicology test systems, mostly as class II devices, are in effect *(66)*. After initial consideration as class III devices, IVDs for measuring levels of cyclosporine and tacrolimus have been down classified by FDA as class II devices, subject to special controls. These special controls are contained in a guidance document *(67)*. The guidance also contains study recommendations used by the agency to evaluate substantial equivalence to a similar legally marketed device.

7.5. Genomic IVD Clearances

The FDA has cleared four IVD 510(k) submissions that incorporate genomics as class II devices. Two of these are for drug-metabolizing enzymes cytochrome P450 2D6 and 2C19. A special control guidance for these is available *(68)*. The third is for the enzyme UGT1A1, which is active in the metabolism of certain drugs such as irinotecan, which is used in colorectal cancer treatment. Summaries of the 510(k)s for these three enzymes are available on the OIVD Web page. The fourth is for a genetic disease, cystic fibrosis *(69)*. Summaries and Guidance documents can provide information on what the FDA evaluates in its review of these types of submissions.

7.6. Drug-Diagnostic Codevelopment

The FDA refers to the process when a new drug is being developed along with an IVD to provide information the physician can use to guide in the patient's therapy with that drug as *codevelopment*. A concept paper has been issued addressing this *(70)*, and a draft guidance is expected to be issued in 2008. Perhaps the best known codevelopment activity is the drug Herceptin (Trastuzumab, Genentech, South San Francisco, CA) and the associated IVD for HER2neu. The FDA approved the NDA for the drug and the PMA for the IVD on the same day.

In the case of a drug under development, where a set of genes is thought to be associated with adverse reaction to the drug, samples can be collected from subjects in clinical trials who experienced toxicity due to drug exposure if the genomic data were not used as part of the exclusion criteria for entering the clinical trial. Aliquoted and appropriately stored samples can be used in the assay validation studies. For example, one set of patient samples is used for the training set and an independent set of samples from other patients is used for cross-validation in typical study designs that use microarray data for class prediction.

8. Conclusion

Controls and metrics need to be applied to many of the major steps involved in the generation of microarray data in order to have confidence in the reproducibility and the validity of the results generated. The steps where quality metrics can be captured and where external controls can be applied are outlined in **Fig. 1**. Universal standards remain an important goal for improving individual laboratory performance, protocol optimization, and methods standardization. For toxicogenomics, additional information is needed on the effect of variations in animal study protocols on gene expression level variance.

As identified in **Section 7**, several guidance documents are available from the FDA on how to prepare medical device submissions and on genomics in general. In addition, there are guidance documents on several pharmacogenomic IVDs the agency has cleared for commercial marketing, as well as summaries of the FDA reviews. In cases where a manufacturer desires to commercially market an IVD for use in toxicogenomics, the commercialization plan should include early contact and dialogue with the FDA.

Acknowledgment

This article represents the professional opinions and statements of the authors and is not an official document, guidance, or policy of the U.S. Government, Department of Health and Human Services (DHHS), or the FDA, nor should any official endorsement be inferred.

References

1. Waring, J.F., Jolly, R.A., Ciurlionis, R., Lum, P.Y., Praestgaard, J.T., Morfitt, D.C., et al. (2001) Clustering of hepatotoxins based on mechanism of toxicity using gene expression profiles. *Toxicol. Appl. Pharmacol* . **175**, 28–42.
2. Thomas, R.S., Rank, D.R., Penn, S.G., Zastrow, G.M., Hayes, K.R., Pande, K., et al. (2001) Identification of toxicologically predictive gene sets using cDNA microarrays. *Mol. Pharmacol.* **60**, 1189–1194.
3. Pennie, W., Pettit, S.D., and Lord, P.G. (2004) Toxicogenomics in Risk Assessment: an overview of an HESI collaborative research program. *Environ. Health Perspect.* **112**, 417–419.
4. Shi, L., Reid, L.H., Jones, W.D., Shippy, R., Warrington, J.A., Baker, S.C. et al. (2006) The MicroArray Quality Control (MAQC) project shows inter- and intraplatform reproducibility of gene expression measurements. *Nat. Biotechnol.* **24**, 1151–1161.
5. Bammler, T., Beyer, R.P., Bhattacharya, S., Boorman, G.A., Boyles, A., Bradford, B.U. et al. (2005) Standardizing global gene expression analysis between laboratories and across platforms. *Nat. Methods* **2**, 351–356.

6. Kuo, W.P., Liu, F., Trimarchi, J., Punzo, C., Lombardi, M., Sarang, J. et al. (2006) A sequence-oriented comparison of gene expression measurements across different hybridization-based technologies. *Nat. Biotechnol.* **24**, 832–840.

7. Hoffman, E.P., Awad, T., Palma, J., Webster, T., Hubbell, E., Warrington, J.A., et al. (2004) Expression profiling – best practices for data generation and interpretation in clinical trials. *Nat. Rev. Genet.* **5**, 229–237.

8. Dumur, C.I., Nasim, S., Best, A.M., Archer, K.J., Ladd, A.C., Mas, V.R., et al. (2004) Evaluation of quality-control criteria for microarray gene expression analysis. *Clin. Chem.* **50**, 1994–2002.

9. Carter, D.E., Robinson, J.F., Allister, E.M., Huff, M.W., and Hegele, R.A. (2005) Quality assessment of microarray experiments. *Clin. Biochem.* **38**, 639–642.

10. Benes, V. and Muckenthaler, M. (2003) Standardization of protocols in cDNA microarray analysis. *TRENDS Biochem. Sci.* **28**, 244–249.

11. External RNA Controls Consortium. (2005) The external RNA controls consortium: a progress report. *Nat. Methods* **2**, 731–734.

12. Thompson, K.L., Rosenzweig, B.A., Pine, P.S., Retief, J., Turpaz, Y., Afshari, C.A., et al. (2005) Use of a mixed tissue RNA design for performance assessments on multiple microarray formats. *Nucleic Acids Res.*, **33**, e187.

13. Mattes, W.B., Pettit, S.D., Sansone, S-A., Bushel, P.R., and Waters, M.D. (2004) Database development in toxicogenomics: issues and efforts. *Environ. Health Perspect.* **112**, 495–505.

14. Reid, L.H. and Casey, S. (2005) The value of a proficiency testing program to monitor performance in microarray laboratories. *Pharm. Disc.* **5**, 20–25.

15. Food and Drug Administration. Table of valid genomic biomarkers in the context of approved drug labels. Available at http://www.fda.gov/cder/genomics/genomic_biomarkers_table.htm.

16. Ganter, B., Tugendreich, S., Pearson, C., Ayanoglu, E., Baumhueter, S., Bostian, K., et al. (2005) Development of a large-scale chemogenomics database to improve drug candidate selection and to understand mechanisms of chemical toxicity and action. *J. Biotechnol.* **119**, 219–244.

17. HESI Committee on the Application of Genomics in Mechanism-Based Risk Assessment, Baseline Animal Data Working Group. Available at http://www.hesiglobal.org/Committees/TechnicalCommittees/Genomics/default.htm.

18. Akhtar, R.A., Reddy, A.B., Maywood, E.S., Clayton, J.D., King, V.M., Smtih, A.G., et al. (2002) Circadian cycling of the mouse liver transcriptome, as revealed by cDNA microarray is driven by the suprachiasmatic nucleus. *Curr. Biol.* **12**, 540–550.

19. Storch, K.F., Lipan, O., Leykin, I., Viswanathan, N., Davis, F.C., Wong, W.H., and Weitz, C.J. (2002) Extensive and divergent circadian gene expression in liver and heart. *Nature* **417**, 78–83.

20. Kita, Y., Shiozawa, M., Jin, W., Majewski, R.R., Besharse, J.C., Greene, A.S., and Jacob, H.J. (2002) Implications of circadian gene expression in kidney, liver and the effects of fasting on pharmacogenomic studies. *Pharmacogenetics* **12**, 55–65.

21. Boorman, G.A., Blackshear, P.E., Parker, J.S., Lobenhofer, E.K., Malarkey, D.E., Vallant, M.K., et al. (2005) Hepatic gene expression changes throughout the day in the Fischer rat: Implications for toxicogenomic experiments. *Toxicol. Sci.* **86**, 185–193.
22. Watson, W.H., Dahm, L.J., and Jones, D.P. (2003) Mechanisms of chemically induced liver disease, In: *Hepatology: A Textbook of Liver Disease*, Vol. II (Zakim, D. and Boyer, T.D., eds.), Saunders, Philadelphia, pp. 739–753.
23. Morgan, K.T., Jayyosi, Z., Hower, M.A., Pino, M.V., Connolly, T.M., Kotlenga, K., et al. (2005) The hepatic transcriptome as a window on whole-body physiology and pathophysiology. *Toxicol. Pathol.* **33,** 136–145.
24. Stokkan, K.A., Yamazaki, S., Tei, H., Sakaki, Y., and Menaker, M. (2001) Entrainment of the circadian clock in the liver by feeding. *Science* **291,** 490–493.
25. Takashima, K., Mizukawa, Y., Morishita, K., Okuyama, M., Kasahara, T., Toritsuka, N., et al. (2006) Effect of the difference in vehicles on gene expression in the rat liver—analysis of the control data in the Toxicogenomics Project Database. *Life Sci.* **78**, 2787–2796.
26. Sakamoto, A., Imai, J., Nishikawa, A., Honma, R., Ito, E., Yanagisawa, Y., et al. (2005) Influence of inhalation anesthesia assessed by comprehensive gene expression profiling. *Gene* **356**, 39–48.
27. Boorman, G.A., Irwin, R.D., Vallant, M.K., Gerken, D.K., Lobenhofer, E.K., Hejtmancik, M.R., et al. (2005) Variation in the hepatic gene expression in individual male Fischer rats. *Toxicol. Pathol.* **33,** 102–110.
28. Whitney, A.R., Diehn, M., Popper, S.J., Alizadeh, A.A., Boldrick, J.C., Relman, D.A., Brown, P.O. (2003) Individuality and variation in gene expression patterns in human blood. *Proc. Natl. Acad. Sci. U. S. A.* **100,** 1896–1901.
29. Waters, M.D. and Fostel, J.M. (2004) Toxicogenomics and systems toxicology: aims and prospects. *Nat. Genet.* **5,** 936–948.
30. Irwin, R.D., Parker, J.S., Lobenhofer, E.K., Burka, L.T., Blackshear, P.E., Vallant, M.K., et al. (2005) Transcriptional profiling of the left and median liver lobes of male f344/n rats following exposure to acetaminophen. *Toxicol. Pathol.* **33**, 111–117.
31. Hamadeh, H.K., Jayadev, S., Gaillard, E.T., Huang, Q., Stoll, R., Blanchard, K., et al. (2004) Integration of clinical and gene expression endpoints to explore furan-mediated hepatotoxicity. *Mutat. Res.* **549**, 169–183.
32. Pine, P., Rosenzweig, B.A., Turpaz, Y., and Thompson, K. (2007) Characterization of alterations in rat liver microarray data induced by tissue handling. BMC Biotechnol. **7**, 57.
33. Srinivasan, M., Sedmak, D., and Jewell, S. (2002) Effect of fixatives and tissue processing on the content and integrity of nucleic acids. *Am. J. Pathol.* **161**, 1961–1971.
34. Lewis, F., Maughan, N.J., Smith, V., Hillan, K., and Quirke, P. (2001) Unlocking the archive—gene expression in paraffin-embedded tissue. *J. Pathol.* **195**, 66–71.
35. Michel, C., Desdouets, C., Sacre-Salem, B., Gautier, J.C., Roberts, R., and Boitier, E. (2003) Liver gene expression profiles of rats treated with clofibric acid: comparison of whole liver and laser capture microdissected liver. *Am. J. Pathol.* **163,** 2191–2199.

36. Luzzi, V., Mahadevappa, M., Raja, R., Warrington, J.A., Watson, M.A. (2003) Accurate and reproducible gene expression profiles from laser capture microdissection, transcript amplification, and high density oligonucleotide microarray analysis. *J. Mol. Diagn.* **5**, 9–14.

37. Baechler, E.C., Batliwalla, F.M., Karypis, G., Gaffney, P.M., Moser, K., Ortmann, W.A., et al. (2004) Expression levels for many genes in human peripheral blood cells are highly sensitive to ex vivo incubation. *Genes Immun.* **5**, 347–353.

38. Debey, S., Schoenbeck, U., Hellmich, M., Gathof, B.S., Pillai, R., Zander, T., and Schultze, J.L. (2004) Comparison of different isolation techniques prior gene expression profiling of blood derived cells: impact on physiological responses, on overall expression and the role of different cell types. *Pharmacogenomics J.* **4**, 193–207.

39. Rainen, L., Oelmueller, U., Jurgensen, S., Wyrich, R., Ballas, C., Schram, J., et al. (2002) Stabilization of mRNA expression in whole blood samples. *Clin. Chem.* **48**, 1883–1890.

40. Affymetrix. Technical Note: Globin reduction protocol: a method for processing whole blood RNA samples for improved array results. Available at http://www.affymetrix.com/support/technical/technotes/blood2_technote.pdf.

41. Lu, Q., Cao, T., Zhang, Z., and Liu, W. (2004) Multiple gene differential expression pattersn in huma ischemic liver: Safe limit of warm ischemic time. *World J. Gastroenterol.* **10**, 2130–2133.

42. Cronin, M., Ghosh, K., Sistare, F., Quackenbush, J., Vilker, V., and O'Connell, C. (2004) Universal RNA reference materials for gene expression. *Clin. Chem.*, **50**, 1464–1471.

43. Imbeaud, S., Graudens, E., Boulanger, V., Barlet, X., Zaborski, P., Eveno, E., et al. (2005) Towards standardization of RNA quality assessment using user-independent classifiers of microcapillary electrophoresis traces. *Nucleic Acids Res.* **33**, e56.

44. Schroeder, A., Mueller, O., Stocker, S., Salowsky, R., Leiber, M., Gassmann, M., et al. (2006) The RIN: an RNA integrity number for assigning integrity values to RNA measurements. *BMC Mol. Biol.* **7**, 3.

45. Auer, H., Lyianarachchi, S., Newsom, D., Klisovic, M.I., Marcucci, G., and Kornacker, K. (2003) Chipping away at the chip bias: RNA degradation in microarray analysis. *Nat. Genet.* **35**, 292–293.

46. Columbus Children's Research Institute, Degradometer v.1.41 software. Available at http://www.dnaarrays.org.

47. Ambion Technical Bulletin. Available at http://www.ambion.com/techlib/tn/83/8311.html.

48. Fare, T.L., Coffey, E.M., Dai, H., He, Y.D., Kessler, D.A., Kilian, K.A., et al. (2003) Effects of atmospheric ozone on microarray data quality. *Anal. Chem.* **75**, 4672–4675.

49. Tong, W., Lucas, A.B., Shippy, R., Fan, X., Fang, H., Hong, H. et al. (2006) Evaluation of external RNA controls for the assessment of microarray performance. *Nat. Biotechnol.* **24**, 1132–1139.

50. Clinical and Laboratory Standards Institute (CLSI). (2006) *Use of External RNA Controls in Gene Expression Assays; Approved Guideline.* CLSI document MM16-A. Wayne, PA, Clinical and Laboratory Standards Institute.

51. Sharov, V., Kwong, K.Y., Frank, B., Chen, E., Hasseman, J., Gaspard, R., et al. (2004) The limits of log-ratios. *BMC Biotechnol.* **4,** 3.

52. ILSI. Report from ILSI Health and Environmental Sciences Institute (HESI) Technical Committee on the Application of Genomics to Mechanism-Based Risk Assessment. Available at http://dels.nas.edu/emergingissues/docs/Pettit.pdf.

53. Finkelstein, D.B. (2005) Trends in the quality of data from 5168 oligonucleotide microarrays from a single facility. *J. Biomol. Techiques* **16,**143–153.

54. Affymetrix. Statistical Applications Description Document. Available at http://www.affymetrix.com/support/technical/whitepapers/sadd_whitepaper.pdf.

55. Jones, L., Goldstein, D.R., Hughes, G., Strand, A.D., Collin, F., Dunnett, S.B., et al.(2006) Assessment of the relationship between pre-chip and post-chip quality measures for Affymetrix GeneChip expression data. *BMC Bioinformatics* **7**, 211.

56. External RNA Controls Consortium. (2005) Proposed methods for testing and selecting the ERCC external RNA controls. *BMC Genomics* **6**, 150.

57. Federal Food Drug and Cosmetic Act. Section 201. 21 United States Code (USC) 321.

58. FDA. Industry Support. Available at http://www.fda.gov/cdrh/industry/support/.

59. 21 CFR 862.9.

60. 21 CFR 807.81, 807.87, 807.92.

61. 21 CFR Part 814.

62. FDA. Office of In Vitro Diagnostic Device Evaluation and Safety. Available at http://www.fda.gov/cdrh/oivd/index.html.

63. FDA. Information on Releasable 510 (k)s Available at http://www.fda.gov/cdrh/510khome.html.

64. 21 CFR 812.2 (c).

65. FDA. Pre-IDE Program: Issues and Answers. Issued March 25, 1999. Available at http://www.fda.gov/cdrh/ode/d99–1.html.

66. 21 CFR 862, Subpart D.

67. FDA. Class II Special Control Guidance Document: Cyclosporine and Tacrolimus Assays; Guidance for Industry and FDA. Issued September 16, 2002. Available at http://www.fda.gov/cdrh/ode/guidance/1380.html.

68. FDA. Guidance for Industry and FDA Staff Class II Special Controls Guidance Document: Drug Metabolizing Enzyme Genotyping System. Issued March 10, 2005. Available at http://www.fda.gov/cdrh/oivd/guidance/1551.html.

69. FDA. System, Cystic Fibrosis Transmembrane Conductance Regulator, Gene Mutation Detection. Available at http://www.accessdata.fda.gov/scripts/cdrh/cfdocs/cfivd/index.cfm?db=PMN&id=K043011.

70. FDA. Drug-Diagnostic Co-development Concept Paper. Issued April 2005. Available at http://www.fda.gov/cder/genomics/pharmacoconceptfn.pdf.

4

Role of Statistics in Toxicogenomics

Michael Elashoff

Summary

In this chapter, we provide a structured approach to the statistical analysis of toxicogenomic data, from the assessment of data quality to data exploration, gene and pathway level analysis, and finally predictive model building. This type of analysis approach can yield toxicogenomic models that provide validated and reliable information about the toxicity of compounds. In addition, we provide study design recommendations for genomic studies in toxicology, covering areas of power, sample size, the need for replicates, and the issue of sample pooling.

Key Words: data quality; multivariate analysis; power; predictive modeling; sample size; statistics; study design.

1. Introduction

In contrast with a classic toxicology study, toxicogenomics profiling of compounds produces vastly larger quantities of data. Each animal sample generates more than 20,000 gene expression values using the latest generation of microarrays, there are typically 30 to 40 animals tested for each compound, and toxicogenomic databases may contain tens or hundreds of compounds. A great variety of gene expression analysis tools have sprung up to provide point-and-click analyses of this large volume of data. However, there is little standardization in analysis methods for toxicogenomic studies, and the numbers of genes involved magnifies the effect of differences in analytic approaches. These points serve to underscore the critical need for rigorous statistical analysis in the toxicogenomics field.

From: *Methods in Molecular Biology, vol. 460: Essential Concepts in Toxicogenomics*
Edited by: D. L. Mendrick and W. B. Mattes © Humana Press, Totowa, NJ

The code needed to run the analyses included in this chapter can be found at http://www.elashoffconsulting.com. It runs in the analysis software "R" with the Bioconductor package. R is an open-source statistical analysis package, based on the same underlying syntax as S-Plus. R/Bioconductor *(1)* is rapidly becoming the standard tool for statistical analysis of microarray data and is available for free at http://www.bioconductor.org.

2. Individual Toxicogenomic Studies

In this section, we discuss the analysis of an individual toxicogenomic study. A typical study design includes several time points, several doses, and a vehicle control group at each time point. Within each time/dose group there are typically three to six animals, with a single RNA sample run for each animal.

After generating the gene expression data, there are some basic questions that one would ask:

- Is the data of sufficient quality to analyze?
- Is there a treatment effect?
- What are the regulated genes/pathways?
- What does it mean in a toxicological context?

This chapter will discuss the statistical considerations that go into answering these questions.

2.1. Assessment of Data Quality

Data quality is a critical component of any genomic analysis. With the multitude of genes measured for each sample, combined with uncertainties regarding the expected expression ranges and regulation for each gene, it may be hard to distinguish interesting biological variability from poor-quality data that should be discarded. Quality metrics play an important role in this determination.

2.1.1. Quality Metrics

Quality control (QC) metrics are intended to be sensitive to variation due to data quality but insensitive to variation due to biological differences in gene expression. For each of the QC metrics to be discussed below, there are three basic approaches to analyzing them:

1. Thresholds: applying a predetermined passing and failing range for the metric (e.g., percent present <30% = fail).
2. Consistency: looking for metric values that are outside of norm within the study (e.g., percent present outside its 95% confidence interval for the study = fail).
3. Balance: comparing the study groups for similarity of metric distributions.

Of these, (2) and (3) are probably the most useful. The challenge with thresholds is that the thresholds are usually chip type and sample type specific and require substantial numbers of chips to appropriately set the thresholds.

2.1.1.1. PERCENT PRESENT

Perhaps the most informative measure of expression data quality is the percent present, or the fraction of genes that are deemed to be present by the chip's analysis package divided by the total number of genes on the chip. As noted above, the interpretation of percent present is chip and sample type dependent, so that absolute thresholds cannot be set. For example, the Affymetrix (Santa Clara, CA) Hu133A chip tends to have approximately twice the percent present value compared with the same sample run on the Hu133B chip. And a percent present value of 40% on the Hu133A would be relatively low for a liver sample but relatively high for a blood sample. Rather, the analysis of percent present focuses on the distribution of values within a particular study (consistency) and between study groups (balance).

2.1.1.2. 5′ 3′ RATIO

Expression arrays generally have one or more control genes on the chip with both a 5′ fragment and a 3′ fragment. For example, Affymetrix chips have GapDH and Beta- Actin. The ratio of the 5′ expression to the 3′ expression is a useful metric for RNA degradation and hence data quality. Additionally, there is a whole chip measure that can be used for Affymetrix chips. This RNA degradation metric makes use of the multiple probes within each probeset and measures the average loss in signal from 3′ probes to 5′ probes.

2.1.1.3. SCALE FACTOR

Affymetrix has used the scale factor as a measure of QC for their chips. Scale factor is proportional to $100/u_{trim}$, where u_{trim} is the trimmed[1] mean of the unnormalized MAS5 expression values. The proportionality is a result of Affymetrix allowing users to define target intensities other than 100. The concept is easily extended to other chips by taking the trimmed mean of expression values using whatever expression measure is appropriate for that platform.

2.1.1.4. AFFYMETRIX SPECIFIC (MM > PM)

Affymetrix expression chips are designed as a series pairs of perfect match (PM) and mismatch (MM) probes. The number of probe pairs in which the PM

---

[1] Affymetrix support bulletin. http://www.affymetrix.com/support/technical/whitepapers/sadd_whitepaper.pdf

is greater than the MM probe has been found to be an informative measure of chip quality.

2.1.1.5. Signal Distribution

Comparison of the expression distributions can reveal data quality differences and may be particularly sensitive to differences at the high and low end of expression that is not captured in other QC metrics.

2.1.1.6. Spike-in Control Genes

In general, spiked-in control genes have not been found to be useful for microarray QC.

2.1.2. Correlation

The preceding quality metrics were mainly focused on general characteristics of the data rather than on the expression values themselves. An assessment of data quality should also include an analysis on the consistency of the gene expression values from sample to sample within the study. For any pair of samples, a common measure of similarity is the Pearson correlation. It can be calculated using any subset of genes, but for data quality purposes it is most informative to use the entire gene set. That concept extends to the correlation matrix, which gives the correlation values of each sample to each other individual sample in the study (see **Table 1** for an example). An important data quality metric can be derived from the correlation matrix by taking the means for each sample. This gives an average correlation value for each sample. If the average correlation for a sample is very low, then its gene expression profile differs from the other samples in the study.

Within an experimental group, average correlation values tend to be .95 or greater. You might expect that having samples in different experimental groups (time/dose) would result in low correlation values when assessed over the whole study, but in practice the differences due to time and dose are relatively modest

Table 1
Correlation Matrix

	Sample 1	Sample 2	Sample 3	Sample 4
Sample 1	1	.80	.95	.96
Sample 2	.80	1	.87	.83
Sample 3	.95	.87	1	.97
Sample 4	.96	.83	.97	1
Average	.90	.83	.93	.92

in the context of a whole genome similarity measure. Even samples treated with relatively strong toxins generally show average correlation values in the range .90 to .95 when compared with untreated controls. As a general rule, an average correlation value below .90 usually signifies that the sample is of low quality compared with the other samples in the study. It is important to note that correlation values must be calculated on logged expression values.

2.2. Data Exploration

2.2.1. Principal Component Analysis

Principal component analysis (PCA) is a multivariate technique for visualizing multidimensional data sets *(2)*. PCA can be based on any set of genes, although it is usually performed on one or more of the following:

- All genes
- Changing genes
- Genes for each pathway

The analysis generates a series of principal components whose values can be plotted against each other. Samples will appear as individual points in these plots.

2.2.1.1. INTERPRETATION

The All Genes PCA will show the predominant patterns in the gene expression data, which may or may not correspond with the study groupings. If poor-quality data or other problems (e.g., wrong sample) were not flagged in the initial QC of the data, these will often show up as outliers in the PCA plot. Differences in sample processing may also appear prominent in the plot. Note that in some cases, the grouping is driven by large numbers of genes, in others by small numbers of highly regulated genes.

In **Example 1**, the main pattern in the expression data is the difference between the high-dose 24-h samples and the remainder of the samples (**Fig. 1**). We will discuss how to investigate the genes that are driving this difference in the following section. The Changing Genes PCA is based on a gene set derived from a statistical comparison of the study groups (see next section). Not surprisingly then, the PCA plot for this gene set will show generally grouping by dose/time groups. Its main utility is to show individual samples that do not fit the pattern (treated animals that group with vehicles or vice versa). These samples may warrant further investigation.

In **Example 2**, by selecting a subset of the genes that show a dose and/or time response, the dose and time groups show segregation in the PCA plot (**Fig. 2**). On the x axis, component 1 seems to represent the dose effect, whereas component 2 on the y axis shows a time effect. The dose response seems more evident at 24 h than at 6 h. Again, though, as we have forced the PCA to show

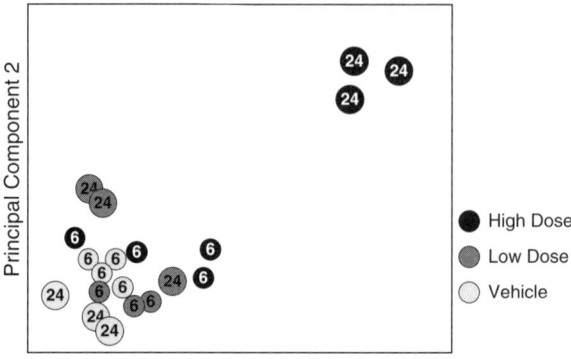

Fig. 1. Principal components plot of individual samples within an example toxicoge-nomics study. Each point represents an animal sample; points are colored by dose and labeled by time (hours). In this example, the 24-h high-dose samples appear to be different from the remainder of the samples.

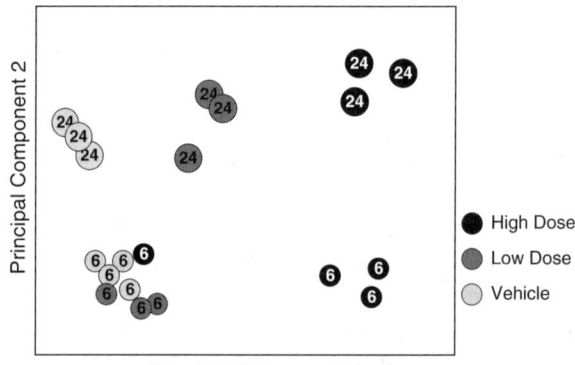

Fig. 2. Principal components plot of individual samples within an example toxicoge-nomics study. Each point represents an animal sample; points are colored by dose and labeled by time (hours). In this example, there appears to be both dose effect and a time effect within the study.

a pattern, we need to interpret the pattern cautiously. We can also see that one of the high-dose 6-h samples is inconsistent with the overall pattern.

2.2.1.2. PERCENT VARIANCE

For each principal component there is a corresponding value representing the percent variance explained by the component. This enables a rough assessment of how reflective the PCA plots of the first several components are of the entire gene expression data set. To put the percent variance into perspective, one can compare the variance to that expected by chance, which is 100/no. samples (e.g., with 25 samples, the percent variation explained by chance would be 4% per component).

2.3. Gene Level Analysis

2.3.1. Fold Change

The basic analysis method in genomics is the fold change (FC) analysis, also called the two-sample *t*-test. It is used when there are two groups of samples, for example high-dose samples and control samples. It statistically tests for a difference in mean logged expression levels between the two groups, giving a *p* value and a fold change (difference of means on the log scale, converted into a ratio of means). It is a basic statistical analysis made complex by the numbers of genes involved and the consequent issue of false positives.

2.3.2. Cutoffs

The question invariably arises as to what *p* value and/or fold change criteria should one use to determine the set of regulated genes. There is no "best" answer to this question, as it depends on many factors that may be context specific. However, we can lay out a general statistical framework for answering the question that can be applied to specific study designs or research questions.

The approach is based on the concept of the false-discovery rate (FDR). This is the ratio of the number of false-positive genes (genes that appear significant but are not truly regulated) to the total of true-positive and false-positive genes. For example, if one performed an analysis of 10,000 genes and found 1000 significant at $p < .05$, we would estimate that 500 of the genes (10,000*.05) would be false positives and thus the FDR would be 50%. Related to the FDR is power, the ratio of the number of true positives over the total number of truly regulated genes. For example, a power of 90% means that the list of significant genes contains 90% of the genes that are truly regulated by the compound. The "best" cutoff for an analysis can then be expressed as the cutoff that achieves the maximal power while minimizing the FDR at some preset level.

Example 1

For the first example, consider a basic analysis with four animals in each of two groups. There are 8800 genes being assayed, and we assume a coefficient of variation (CV) of 30% for the gene expression measurements. We will also need to postulate some statistics regarding the degree of gene regulation, in this case we assume 500 regulated genes out of the 8800 with fold changes in the range of 1.25 to 3.0.

Assumptions:

- Two groups: treated versus control, $N = 4$/group
- 8800 genes (Rgu34a chip)
- CV = 30%
- 500 regulated genes (100 at 1.25, 100 at 1.5, 100 at 1.75, 100 at 2, 100 at 3)

Based on these assumptions, we can calculate the FDR and the power for a range of p value and fold change cutoffs. For this study, we may want to ensure that the FDR is less than 5%. The analysis cutoffs that are compatible with a FDR <5% are highlighted in **Table 2**. We can then choose the analysis criteria that maximize the power within the highlighted area. In this case, a $p < .05$, FC > 1.7 cutoff gives the highest power (53%) (**Table 2**).

Example 2

To illustrate how the analysis criteria need to be determined on a case-by-case basis, consider a second example. In this example, we expect the same number of regulated genes (500), but at a lower fold change (1.25 to 1.5), and we have an increased sample size for this study to detect the more subtle expression changes that are expected (**Table 3**).

Assumptions:

- Two groups: treated versus control, $N = 8$/group
- 8800 genes (Rgu34a chip)
- CV = 30%
- 500 regulated genes (250 at 1.25, 250 at 1.5, 0 at 1.75, 0 at 2, 0 at 3)

Table 2
Example 1 FDR and Power

	FC cutoff (%)				
p-value cutoff	>1	>1.3	>1.5	>1.7	>2
<.001	5.8, 28	4.7, 28	1.7, 28	0.2, 27	<0.1, 23
<.005	15, 47	12, 47	4.1, 45	0.5, 42	<0.1, 31
<.01	23, 56	19, 60	5.5, 54	0.5, 47	<0.1, 33
<.05	52, 76	40, 75	8.6, 67	0.6, 53	<0.1, 35

Table 3
Example 2 FDR and Power

p-value cutoff	FC cutoff (%)				
	>1	>1.3	>1.5	>1.7	>2
<.001	5.8, 28	4.5, 27	.3, 21	<.1, 5	<.1, .1
<.005	15, 46	10, 45	.3, 26	<.1, 5	<.1, .1
<.01	24, 54	13, 51	.3, 26	<.1, 6	<.1, .1
<.05	52, 75	20, 62	.4, 27	<.1, 6	<.1, .1

The same p value and fold change filter used for **Example 1** (FC > 1.7, $p < .05$) would here produce a very low power (6%). But by using a FC > 1.5, $p < .05$ filter, we can increase the power to 27%. And it turns out that by doing further optimization, one can achieve a power of 45% while still maintaining a FDR <5% (FC > 1.4, $p < .05$).

In the two examples, we used a FDR threshold of 5%, meaning that we wanted no more than 5% of the genes in the list of significant genes to be false positives. But the context may dictate higher or lower FDR requirements. For example, if one is interested in fully characterizing a compound's effects, it may be advantageous to be more complete in the determination of the regulated gene, at the expense of some additional false positives.

2.3.3. Filtering

Typical studies will have only about half of the genes called present by the image processing algorithms. One way to reduce the false-positive rate/false-discovery rate is to remove genes from the analysis that are not called present. This is certainly defensible from a statistical point of view. A caveat, though, is that this process may also remove genes that are actually present but expressed at low levels. For example, in the Affymetrix Latin Square study (http://www.affymetrix.com), genes at a known concentration of 3 pM were called present >95% of the time, but as concentration dropped to 1 pM, the present frequency was 75% and at 0.33 pM the present frequency was ~35%. Thus, if one is interested in low-expressing genes, a filter based on present calls may be counterproductive.

2.3.4. Analysis of Variance

Typically, a toxicogenomic study will have multiple doses and/or time points. In these cases, rather than performing a series of pairwise fold change analyses (e.g., high-dose vs. control at each time point), it is preferable to perform an analysis of variance (ANOVA). An ANOVA model can account for time effects and dose effects simultaneously. For each gene, an analysis of variance will yield:

- Overall dose effect: tests whether the gene's expression level varies across dose groups.
- Overall time effect: tests whether the gene's expression level varies across time.

2.3.5. Pathway Analysis

With a set of regulated genes, one can start to look across the set for common characteristics. Pathway membership is one such characteristic. A basic measure of the degree to which a pathway is affected by a compound is

$$\frac{\text{No. genes in the pathway that are regulated/No. genes in the pathway}}{\text{No. genes not in the pathway that are regulated/No. genes not in the pathway}}$$

This measure accounts for the fact that pathways have differing numbers of genes. Note that measures relying on counts of genes, such as this, essentially treat genes as independent. But genes are not independent; in fact it is not uncommon for multiple regions of a gene to be present on a given array. To account for the gene-to-gene dependence, one needs to use a measure that estimates and then incorporates the correlation between genes within a pathway. To do this properly, however, requires more samples than the number of genes in the pathway, making this approach infeasible for all but the smaller pathways. In **Section 3.3**, we discuss this approach in the context of a database of compounds, where the sample sizes allow for such an approach.

2.4. What Does It Mean?

Statistical analysis can get to a set of regulated genes/pathways associated with the compound under study. Done properly, one can have a high degree of confidence in those genes. But what do those genes imply about the compound's toxicity? This topic is addressed from a toxicology perspective in detail in the other chapters of this book and from a statistical perspective in the following section.

3. Toxicogenomic Databases

3.1. Assessment of Data Comparability

As discussed in the previous section, the analysis of an individual compound or study can only get so far. To place the expression profile of a compound into context, one needs to examine the profiles of other similar and dissimilar compounds. The power of a large database can only be leveraged if the samples are generally comparable with the samples in the database. Factors to consider include:

- Vehicle controls
- Sex/strain of animals
- Sample processing methods
- Chip processing methods

3.1.1. Normalization

Based on experience, unnormalized data from one site are generally not comparable with unnormalized data from another site. The differences may be caused by one or more of the factors listed above. However, within-study normalization (using the study vehicle treated animals) can often remove much of the cross-study differences while preserving the underlying biological responses. This is because whereas the factors alter the baseline expression levels of genes, they typically do not alter the expression regulation that a compound induces. In statistical terms, this means that the treatment by factor interaction is generally small. For example, differences in labeling reagents may result in substantial differences in gene expression, but these would not be expected to alter the animal's response to treatment and could thus be normalized out of the data. Empirically, sex and strain appear to act in the same manner.

3.1.2. Comparability/Consistency with Database

Following within-study normalization, one should then assess the data comparability between the study samples and the database samples. The methods described in **Section 2.1** and **Section 2.2** can be used for this assessment:

- QC metrics: comparison of metric distribution for the study samples and the database samples.
- PCA: are the study samples grouping with the database samples on the first several principal components.
- Clustering: are the study samples grouping with the database with an acceptable degree of correlation.

Note that these methods represent a necessary but not sufficient condition to establish comparability, as they may not address the biological applicability of comparisons across the database.

3.2. Data Exploration

Clustering and PCA can give a global view of the database, although the interpretation of the patterns can be difficult. The reason is that compounds may group/cluster together for a variety of reasons:

1. Mechanism-of-toxicity effects of compounds. Often these are the investigational target of the study, but they might not represent the majority of the gene expression changes in the study.
2. Degree of toxicity. Beyond a certain level of toxicity, compounds may group together even when the mechanisms are quite different.
3. Non–toxicity effects of compounds. Many compounds will induce gene expression changes that are unrelated to toxicity. For example, these effects may be due to mechanism of action, effects on metabolism, and so forth.
4. Study effects. When clustering based on control-group normalized (fold change) expression data, compounds may group together if they share a control group, as the shared control group introduces a statistical correlation between the compounds. Even when the studies do not share a control group, they may share other study factors that cause the compounds to cluster together. For example, if studies are performed over a period of time, you may observe temporal groupings in the data that could reflect changes in study conduct, sample processing, and so forth.

3.2.1. Clustering Algorithms

Questions often arise as to the best types of clustering algorithm to use. In general, the biological questions of interpretation described in this section outweigh the statistical considerations of clustering methodology.

3.3. Gene Level Analysis

The same types of gene and pathway analyses that can be done on individual compounds (described in **Section 2**) can be performed on sets of compounds. With more samples, though, one can perform more sophisticated versions of these analyses. Note that code for these analyses are available online at the author's Web site.

3.3.1. ANOVA with Study as a Factor

An ANOVA model that includes study, dose, and time as factors can potentially detect subtle effects that would be missed by within-study analyses. The cross-study ANOVA can be performed on the absolute expression values or relative expression values normalized to the study specific controls.

3.3.2. Gene Similarity Analysis

While there are published sets of markers for certain types of toxicities, little is known about the toxicologic significance of the substantial majority of genes. For that reason, it can be of interest to identify genes that act similarly to known toxicity markers based on their gene expression profiles. One way to identify similar genes is the statistical algorithm MatchX *(3)*. The algorithm generates a distance score between the gene of known function and each of the other genes on the array.

3.3.3. Pathway Fold Change

A fold change (*t*-test) analysis compares the means of a single gene for two groups of samples, for example toxic compounds versus nontoxic compounds. One can extend the *t*-test to the multivariate setting and compare the means of a set of genes between two groups of samples. An important use of this method is for pathway analysis, where the test can be used to establish which pathways are differentially regulated. The real advantage to this approach, unlike that described in **Section 2.3** (counting the number of regulated genes in the pathway), is that the correlation of genes within a pathway can be taken into account.

3.4. Predictive Modeling

Predictive modeling attempts to link the gene expression profiles of multiple compounds to an expected behavior of the compound when used in humans. We start this section with a discussion of what is typically thought of as the final phase of the model building process: model validation. This is because how one validates a model drives, in large part, how best to build the model.

3.4.1. Model Validation

Model validation is simple in theory but difficult to rigorously apply in practice. To validate a predictive model, one divides the compounds into a training set and a test (holdout) set. The model is built on the training set and then evaluated using the holdout set. In general, accuracy rates should be reported from the test set only, and one should be suspicious of success rates based on the training set. It is important that the training and test sets are divided by compound and not randomly by sample. This prevents the model from using compound-specific genes to artificially inflate the model success rates.

In developing a predictive model, or in critically assessing a published model, it is important to consider how rigorously the training set and test set are separated. Consider the following examples, based on a hypothetical data set of 50 toxic compounds and 50 non-toxic compounds:

(A) A model is reported that was built by selecting genes with $p < .01$, FC > 1.3 between toxins and controls, and applying a PCA based model with eight components to the training set (35 toxins and 35 controls). It was then applied to the test set and achieved 90% accuracy (27/30 test compounds predicted correctly). It is reported that the test compounds were not used to in the gene selection procedure or in the PCA estimation procedure.

(B) Data are divided into a training set (35 toxins and 35 controls) and a test set (30 compounds). Model optimization was performed where different combinations of p-value filter, FC cutoff, prediction algorithm, and algorithm parameters were assessed. For each of these options, five choices were assessed (e.g., FC cutoff of 1.0, 1.3, 1.5, 1.7, 2.0). The total numbers of models built was $5^4 = 625$. The success rates for these models on the test set ranged from 12/30 (40%) to 27/30 (90%). The best model was identified ($p < .01$, FC > 1.3, PCA, eight components) and will be used to screen compounds.

Clearly, on the basis of what is reported, we would have much less confidence in model B than in model A. However, one might reasonably suspect that model A was actually developed along the lines of model B, and we should be skeptical of both models. What this means is that it is as important to know what was tried and did not work as what was tried and did work. For example, our impression of model B would change if the success rate range across models was 25/30 to 29/30, and the 27/30 model chosen was selected based on a predefined criterion of having the fewest genes in the model.

3.4.2. Model Building

It is an established statistical principle that one cannot rely on training set accuracy to train the model and determine parameters, weights, and so forth. But, as the above section indicated, if one relies on the test set accuracy to establish modeling parameters, the results can be equally suspect if the test is overused.

A reasonable question, in light of this, is how should one set model parameters if not by training or test set success rates. In many cases, in addition to the training/test set, a third set of compound data is required for a reliable estimate of model accuracy: an external validation set. This is a set that stands completely separate from the model building process and is only used once, after the final model is built, to establish model success rates.

As a practical matter, one may not have enough compounds to have three large enough data sets. A valid alternative is to apply compound drop model validation. This uses just two sets: a combined training/test set and an external set. In compound drop, one uses the training/test set, holds out a compound, builds the predictive model, on the remaining set and predicts the holdout compound. The process is repeated for all of the compounds in the set. Model parameters are set based on aggregating the results for each dropped compound. Final model success is then calculated on the external set. What to do if the model does not work on the external validation set is left as an exercise for the reader.

3.4.3. Gene Selection

A variety of gene selection methods have been applied to toxicogenomic modeling. These can generally be grouped into four categories

1. Biological selection: approaches that consider the known genes/pathways involved with toxicity.
2. Gene aggregates: approaches that select "meta" genes such as principal components; these meta genes are then used in the predictive model.
3. Discriminant selection: approaches that consider the genes individually on the basis of how they distinguish toxicity (e.g., fold change analysis).
4. Model selection: approaches that consider how the gene helps the predictive model, generally a multivariate approach in the context of other genes (logistic regression, classification trees).

From a statistical point of view, one can think of the approaches being ordered by how effective they are for model building (worst to best: 1, 2, 3, 4) or by how likely they are to produce models that are not overfit (worst to best: 4, 3, 2, 1). There is no established best approach for toxicogenomic modeling at this time; in practice, any of the four could be used successfully in the context of a statistical rigorous model.

Table 4
Pros and Cons of Classification Methods

Method	Fitting capability	Tendency to overfit	Isolate gene contributions
Clustering	Low	Low	No
Classification tree	Low-medium	Medium	Yes
Logistic regression	Medium	Medium	Yes
K means	Medium	Low	No
Partial least squares	High	Medium	Yes
Support Vector Machine (SVM)	High	High	No
Neural networks	High	High	No
Discriminant analysis	Medium	Low	No

3.4.4. Classification Methods

A variety of classification methods have been applied to toxicogenomic modeling. An in-depth description of the individual methods is beyond the scope of this chapter, but an overview of the pros and cons of the methods is shown in **Table 4** *(4)*.

4. Study Design for Toxicogenomics
4.1. Time Points

If one knew the best time point for a given compound, running additional time points would generally be of little added benefit for the analysis of the gene expression data. This is a result of the fact that animals are sacrificed at the time of RNA collection, and so we cannot observe a time course of expression response within an animal even with a multiple time point study (although this will change as blood toxicogenomic profiling becomes more feasible). However, one may not know the best time point for a compound, particularly an investigational compound, and so three or four time points are generally used for a study. The best choice of time points has not been extensively studied.

4.1.1. Doses

From a statistical point of view, the more dose groups the better. Using multiple dose groups enables a dose-response analysis, not only for overall toxicity but also on the gene level for each individual gene and pathway.

4.1.2. Sample Size

Sample size calculations reflect the trade-off between wanting to maximize the scientific information from the study while minimizing the cost of the study. The calculations must also make certain assumptions about the data, such as degree of gene regulation, biological variability between samples, and so forth. The following sample size conclusions should therefore be viewed as general recommendations rather than absolute requirements. For the analysis, we make the following assumptions:

- Rat chip with 30,000 genes
- 1000 regulated genes (3.3% of total gene number)
- Fold change range (high vs. vehicle) = 2.0 for regulated genes
- Coefficient of variation (CV) = 30%
- $p < .001$ criteria for significant genes

   (A) Three time points and three dose groups (vehicle, low, high), with an equal number of animals per time/dose group as shown in **Table 5**.

Table 5
Study Design A

Sample size per group (total)	FP rate(%)	TP rate (power)(%)	FDR(%)
2 (18)	~.1	33	12
3 (27)	~.1	76	6
4 (36)	~.1	94	3
5 (45)	~.1	98	3
6 (54)	~.1	99	3

(B) Two time points and two dose groups (vehicle, high), with an equal number of animals per time/dose group as shown in **Table 6**.
(C) One time point and five dose groups, with an equal number of animals per dose group as shown in **Table 7**.

4.1.3. Technical/Biological Replicates

In almost every situation, biological replicates are preferable to technical replicates for drawing conclusions for a study. An analogy can be made to

Table 6
Study Design B

Sample size per group (total)	FP rate(%)	TP rate (power)(%)	FDR(%)
2 (8)	~.1	8	22
3 (12)	~.1	36	14
4 (16)	~.1	68	5
5 (20)	~.1	88	5
6 (24)	~.1	95	3
7 (28)	~.1	98	3
8 (32)	~.1	99	3

Table 7
Study Design C

Sample size per group (total)	FP rate(%)	TP rate (power)(%)	FDR(%)
2 (10)	~.1	28	10
3 (15)	~.1	69	5
4 (20)	~.1	90	4
5 (25)	~.1	96	3
6 (30)	~.1	99	3

clinical trials. Consider a study of a weight-loss drug versus a control. The primary end point of the trial is weight after 6 months. Trial A compares 50 patients on the drug with 50 patients on control, with each patient having a single 6-month weight measurement. Trial B compares one patient on the drug with one patient on the control arm, with each patient weighed 50 times at 6 months. These trials clearly have very different interpretations. The difference comes in our ability to draw conclusions from the analysis that are applicable outside the context of the study. We typically do not care about a specific individual's response to treatment but about an average response to treatment, and the variation in that average, and only trial A gives that information.

There are, of course, intermediates between the two designs, such as using 25 versus 25 patients, each measured twice, and so forth. In assessing this trade-off, an important consideration is the ratio of biological variability to technical variability in the data. The utility of technical replicates will go up as this ratio goes down. In other words, as the scale in our clinical experiment gets more variable relative to the biological differences in weights, it is more important to take multiple measurements for each person. Now, technical variation in gene expression is substantially higher compared with measures such as weight. But so are gene expression responses (biological variation). In fact, biological variations tends to be a factor of two- to fourfold times the technical variation in gene expression studies. Thus, the balance remains shifted toward a trial A design.

An exception to this general rule is the case where one cares a lot about a specific individual. This might be the case in a diagnostic situation, where a decision will be made for an individual patient based on gene expression measurements. In this case, it makes sense to make use of technical replicates to better quantify the gene expression profile for the individual.

4.1.4. Pooled/Unpooled Samples

The considerations to pool or not pool samples are analogous to the considerations for biological versus technical replicates. That discussion centered on the fact that, in order to draw meaningful conclusions, we need to have information on the average response and on the variation in that average across a set of individuals. By definition, a study composed of technical replicates does not yield an average response. Whereas a study with pooled samples does give the average response, it does not give the variation in that response. We do not know whether a gene expression response in a pooled sample is due to a single large response in one individual or to a consistent smaller response across all individuals. For this reason, the use of pooled samples precludes a meaningful analysis of the study data.

References

1. Gentleman, R., Carey, V., Bates, D., Bolstad, B., Dettling, M., Dudait, S. et. al. (2004) Bioconductor: open software development for computational biology and bioinformatics. *Genome Biol.* **5**, R80.
2. Joliffe, I.T. (2002) *Principal Component Analysis*. Springer, New York.
3. Coberley, C., Elashoff, M., and Mertz, L. (2004) Match/X, a gene expression pattern recognition algorithm used to identify genes which may be related to CDC2 function and cell cycle regulation. *Cell Cycle* **3**, 804–810.
4. Hastie, T., Tibshirani, R., and Friedman, J. (2003) *The Elements of Statistical Learning*. Springer, New York

5

Predictive Toxicogenomics in Preclinical Discovery

Scott A. Barros and Rory B. Martin

Summary

The failure of drug candidates during clinical trials due to toxicity, especially hepato-toxicity, is an important and continuing problem in the pharmaceutical industry.

This chapter explores new predictive toxicogenomics approaches to better understand the hepatotoxic potential of human drug candidates and to assess their toxicity earlier in the drug development process. The underlying data consisted of two commercial knowl-edgebases that employed a hybrid experimental design in which human drug-toxicity infor-mation was extracted from the literature, dichotomized, and merged with rat-based gene expression measures (primary isolated hepatocytes and whole liver). Toxicity classification rules were built using a stochastic gradient boosting machine learner, with classification error estimated using a modified bootstrap estimate of true error. Several types of clustering methods were also applied, based on sets of compounds and genes. Robust classification rules were constructed for both *in vitro* (hepatocytes) and *in vivo* (liver) data, based on a high-dose, 24-h design. There appeared to be little overlap between the two classifiers, at least in terms of their gene lists. Robust classifiers could not be fitted when earlier time points and/or low-dose data were included, indicating that experimental design is important for these systems. Our results suggest development of a compound screening assay based on these toxicity classifiers appears feasible, with classifier operating characteristics used to tune a screen for a specific implementation within preclinical testing paradigms.

Key Words: classification rule; *in vitro*; *in vivo*; machine learning; stochastic gradient boosting; toxicity; toxicogenomics.

1. Introduction

The current cost for discovery and development of new drug candidates continues to increase at an alarming rate. Recent industry reviews have estimated the total cost to bring a drug candidate to market rapidly approaching

From: *Methods in Molecular Biology, vol. 460: Essential Concepts in Toxicogenomics*
Edited by: D. L. Mendrick and W. B. Mattes © Humana Press, Totowa, NJ

US$900 million, with an average success rate of 11% (first-in-man to registration) across nine therapeutic areas *(1,2)*. In addition, a comparison of the reasons for drug candidate failures in 1991 versus 2000 revealed that failures due to clinical side effects and nonclinical toxicologic findings had increased roughly 30% and 60%, respectively *(1)*. Although the reasons for this trend may be multifactorial, clearly there is a need to improve toxicology testing paradigms within the industry. It is understood that the standard practice of testing drug candidates at increasing multiples of the anticipated human therapeutic dose in two species (one rodent, one non-rodent) for at least 14 days of repeat dosing is fallible and not entirely predictive of all human acute, chronic, or idiosyncratic toxic outcomes. The obvious challenge and value proposition to the industry is to develop more robust predictive *in vitro* and *in vivo* models to identify toxic liabilities earlier during discovery phases.

For the discovery toxicologist specifically, the objective is to develop tools that enhance the existing activities of the medicinal chemist and pharmacologist during hit-to-lead and early lead optimization stages. Typically, compounds within chemical series are rank-ordered based on physical-chemical properties, potency to target (receptor occupancy, target fidelity, binding characteristics, cytotoxicity, etc.), pharmacokinetics (PK), pharmacodynamics (PD), and efficacy in disease models. To leverage toxicology among these early activities, the ideal assay must (a) provide a robust prediction of toxic potential for subsequent *in vivo* studies, (b) provide a quantitative output for rank-ordering, (c) be compatible with existing screening throughput, (d) be cost effective, and (e) potentially discriminate among toxicities that are or are not due to an intended pharmacologic mechanism. This last point is important on two levels. First, from a philosophical perspective, developing the *n*th assay for rank-ordering of compounds alone is unlikely to be met with enthusiasm from chemists and pharmacologists which already have a substantial stable of quantitative assays at their disposal—a novel assay must provide additional useful information. Second, it would be invaluable to predict chemistry-based toxic potential, as it may be possible to synthetically design around the liability on the lead chemical scaffold. Alternatively, if the toxic potential is predicted to be mechanism-based, project teams could focus on the more traditional route of mitigating toxicities based on exposure and efficacy rather than synthetic design.

Predictive *in vitro* toxicogenomics is one of the most intriguing emerging technologies on the current preclinical toxicology landscape. For this chapter, toxicogenomics is defined as the science of using gene expression profiling from xenobiotic-treated cells or tissues to classify and/or predict various toxic outcomes.

Drug-induced hepatotoxicity accounts for approximately half the cases of acute liver failure in the United States (intentional and unintentional) and is a leading cause of attrition in clinical phases of development *(3,4)*. In addition, postmarketing incidence of drug-induced hepatotoxicity is a leading cause of withdrawals, labeling changes, and black box warnings by regulatory agencies *(4)*. In this context, a toxicogenomic assay designed to predict for liver injury would be of great value. This chapter describes our internal efforts to develop and evaluate predictive *in vitro* and *in vivo* toxicogenomic models for hepatotoxicity based on the raw microarray data within the ToxExpress® database (Gene Logic, Inc., Gaithersburg, MD), with the goal of leveraging this technology within our existing toxicology testing paradigm. Although the *in vitro* and *in vivo* models were developed and evaluated simultaneously, our belief is that implementing toxicogenomics from *in vitro* systems will bear more impact on reducing development attrition and improving the quality of our candidate pipeline.

2. Approach

A classic approach to studying hepatotoxicity is to use rat-based systems; for example, measuring gene expression in rat cells or tissues together with related microscopy, biochemistry, clinical pathology, and histopathology. The utility of this approach rests on the assumption that rat is a good surrogate for human. A second way of leveraging a rat-based design is to incorporate human end points from the start, requiring that the classifier predict for *human* and not rat hepatotoxicity *(5,6)*. This approach also assumes rat to be a good surrogate for human, but necessitates a hybrid experimental design that merges rat-based gene expression with a response culled from human-based literature. This second requirement is nontrivial: it demands a nonstandard experimental design in which observed response is not measured in the same experiment as are inputs. This is a challenging concept and key to understanding and evaluating toxicogenomics as a technology and its potential to be predictive of human outcomes. In the current work, *inputs* refers to gene expression data collected from rat-based systems treated with a reference set of compounds. The *response* refers to the qualitative, literature-based, classification of the behavior of those same reference compounds in humans; that is, *hepatotoxic* or *nonhepatotoxic* (**Fig. 1**).

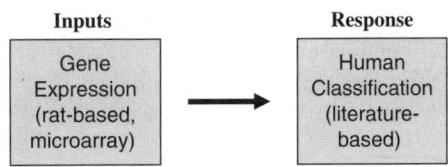

Fig. 1. Toxicogenomics system under study.

Any changes made to the inputs (i.e., experimental design, liver vs. hepato-cytes) do not affect the literature-based classification of the compounds (response). This is a violation of assumptions normally required in machine learning, namely that response must directly depend upon experimental inputs. Though a legitimate cause for concern, this paradigm does confer certain advantages.

First, the hybrid design leverages the considerable existing human literature on compound toxicity, providing confidence during the subjective process of choosing and classifying compounds based on potential to cause hepatocellular injury. Second, experimental duration is drastically reduced, relying on the hypothesis that toxicant-induced gene expression changes manifest themselves far earlier than the pathologies that they eventually precipitate *(5)*. This is key, as traditional *in vivo* toxicology studies are expensive, time consuming, require copious amounts of test article, and are thus ill-suited to the evaluation of large numbers of early-stage compounds.

A number of biological assumptions are inherent in this type of hybrid design. First, we assume that compounds that cause toxicity in humans cause transcriptional changes in rats; whether the same changes are present in rats and humans is an open question. Second, we assume these early gene expression changes foreshadow the later manifestations of hepatotoxicity that form the basis of our literature binned response *(5)*.[1] Before we describe the experimental data used from the ToxExpress database, we review some mathematical terminology and concepts.

3. Mathematical Background

3.1. Terminology

Like many emerging fields, machine learning suffers from a multiplicity of terminology. In the summary below, we list alternative terms for common concepts in the hope that it may be useful to researchers when comparing and integrating results from different sources. In each case, the first listed term is the one normally used in the text:

- Response, outcome, output
- Input, covariate, variable, explanatory variable, predictor
- Binary response, dichotomous response, Boolean response
- Quantitative response, continuous response
- Prediction rule, classifier (for binary response), model

---

[1] For the remainder of this chapter, the uses of *toxic* and *nontoxic* refer specifically to hepatotoxicity, unless otherwise stated.

- Training sample, learning sample
- Test sample, validation sample

3.2. Assessing Performance

Predictive performance has been described as a guiding principle for selecting a prediction rule *(7)*. Not only does it quantify our degree of belief that conclusions based on a model are robust and replicable, but also it facilitates comparisons between different hypotheses (e.g., interactions present/absent), different machine learning methods (e.g., one tree vs. an ensemble of them), and different experimental designs. Being able to reliably quantify confidence in a prediction rule is very important as one of the potential pitfalls of machine learners is they *always* generate a prediction rule.

A theoretical ideal for measuring predictive performance is the true error rate *(8)* of the prediction rule. Given a sample S drawn from some distribution F, with inputs x and observed response y, the true error rate of a prediction rule $M_S(\cdot)$ fitted to S is

$$\text{Expectation } \{Err(y, M_S(x))\}, \text{ over } (x, y) \text{ drawn from } F$$

where $M_S(x)$ is a predicted response, and $Err(\cdot)$ is an error function comparing observed versus predicted response. Examples include 0-1 loss for binary response and squared error for continuous response *(9)*. The above equation is a conditional generalization error, conditioned upon having chosen S as the training sample *(8)*.

In practical terms, true error is not a directly useful quantity, as the very large and independent sample it demands is rarely if ever available in real-world studies. However, a number of estimators are available that seek to approximate true error—these include the bootstrap, split sample, and resubstitution. We now describe each method briefly and compare their strengths and weaknesses (see Ref. *8* for theoretical details).

The resubstitution error rate (sometimes called apparent error) is formed by taking the sample S and dropping it back through the rule $M_S(\cdot)$ that was fitted to it. Resubstitution is highly attractive because of its economy and simplicity: 100% of the data are used for model training, and there are no tuning parameters involved constructing the error estimator. Its Achilles' heel is its downwards bias as an estimator of true error, causing it to paint an overly optimistic picture of predictive performance.

Split sample is a commonly used technique in which a single randomized partition is used to divide the sample into two parts: a training sample used to fit the rule and a test sample with which the rule's performance is quantified.

In our experience, split sample has very wide appeal and usage among all types of scientists. First, it is quick to run. Second, there is maximal clarity regarding the role of data: each datum is used either to train the rule or to test it. Third and possibly most importantly, the method is perceived as impartial, as the test sample (once chosen) can be kept under lock and key.

Completing our offering of sampling methods is the bootstrap *(8,10)*. The name was taken from *The Adventures of Baron Munchausen*, a novel written by 18th century author Rudolph Raspe: "The Baron had fallen to the bottom of a deep lake. Just when it looked like all was lost, he thought to pick himself up by his own bootstraps."

The unusual way in which the fictional Baron avoids drowning—by lifting his own person out of the water—encapsulates a key idea for the mathematical bootstrap, namely self-referentiality. The bootstrap forms an empirical distribution—a probability distribution formed directly from a sample's data points rather than specified as a mathematical function—by placing a probability mass of $1/n$ on each of the n observations in sample S, then uses this empirical distribution as a surrogate for the underlying but unknown population distribution F. Observations are drawn with replacement to create a replicate sample S_i and the process repeated 100 or so times to form a bootstrap ensemble. It is sometimes described as a kind of nonparametric maximum likelihood approach.

Despite its theoretical attraction and published studies establishing its performance versus other sampling methods (see, e.g., Refs. *8,10*) the bootstrap is often viewed with suspicion. Aside from the fact it is more complicated to implement and takes much greater computational effort to run, each observation plays dual roles in a bootstrap, being in turn a member of both training and testing samples. In layman's terms, a bootstrap appears to some to be a free lunch.

Furthermore, we normally perform a finesse when constructing a bootstrap estimator of true error. For resubstitution, the *final* model (rule) that we choose for future prediction and for extracting knowledge about system behavior is the same model we use to estimate error. In contrast, the ensemble of models created during bootstrap are fitted afresh and act as a surrogate for the final model fitted to the entire sample. Of course, the same kind of finesse is also used for the split sample approach.

Differences between these approaches are multifactorial, with it being possible to resample once, repeatedly, or not at all, to draw observations with replacement or without, and to put various degrees of effort on training a model versus testing it. Furthermore, some researchers perceive that advantages of simpler methods such as impartiality have been lost, while it is not clear what has been gained in return.

To study these issues, we recently conducted a Monte Carlo simulation study to illustrate and compare different methods for estimating prediction error, with the aim of providing practical guidance to users of machine learning techniques *(11)*. We used stochastic gradient boosting as a learning algorithm (described more fully in the next section) and considered data from two studies for which the underlying data mechanism was known to be complex: *(1)* a library of 6000 tripeptide substrates collected for analysis of proteasome inhibition as part of anticancer drug design, and *(2)* a cardiovascular study involving 600 subjects receiving antiplatelet treatment for acute coronary syndrome.

We found important differences in the accuracy of the various error estimators examined. Error estimators for split sample and resubstitution, though being the most transparent in action and the simplest to apply, did not quantify the performance of prediction rules as precisely as the bootstrap. In all cases studied, the bootstrap estimator had smaller variance and smaller bias (or nearly so). Distributions for the various error estimators are plotted in **Fig. 2** (*see* Color Plate 4).

Other error estimators exist besides the ones listed above, in particular cross-validation approaches with or without repeated choice of the randomized partition. Although cross-validation with randomized partition was not examined in our study, our expectation is that it would perform similarly to bootstrap and outperform the simpler split sample approach.

Error estimation is only useful because it helps quantify our degree of belief in a final model, the model that we use for future prediction and for

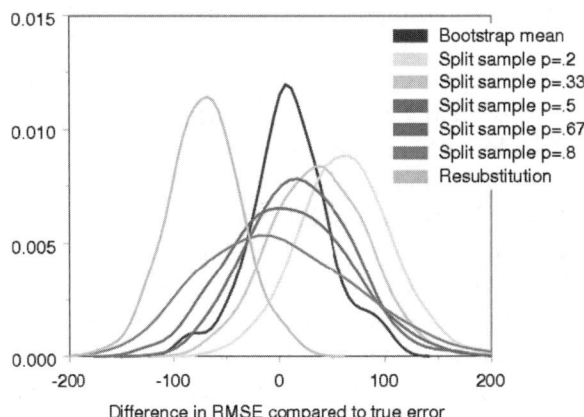

Fig. 2. Distributions for error estimators based on proteasome data. The graphs plot the difference between the estimated error and the true error. Parameter p is the proportion of a baseline sample used for training during error estimation. (*see* Color Plate 4).

understanding a system's behavior. Such a final model is normally fitted to the entire baseline sample. However, some researchers employ a split sample approach from the very beginning, reserving a portion of data solely for model testing. This is akin to strategies employed in sequential clinical trials and it gives maximal confidence that there is no "cheating" involved. However, the price paid is high. When training of the *final* model is limited to a subset of the baseline data, model performance will be significantly worse than that of a model trained using the full, baseline data, a consequence of loss of sample size.

4. Experimental Design

4.1. Experimental Overview

Samples for the *in vitro* and *in vivo* studies were subsets of the primary rat hepatocyte and rat liver portions of the ToxExpress database. To briefly summarize the experimental designs used by Gene Logic, Inc., freshly plated primary hepatocytes from male Sprague-Dawley rats, 6 to 8 weeks of age, were obtained from In Vitro Technologies (Baltimore, MD) and/or Cambrex Bio Science (Baltimore, MD) and incubated overnight in serum-free media (~48 h total post-isolation). The cultures were treated and RNA harvested at 3-, 6-, and 24-h time points for subsequent analysis by microarray. For *in vivo* studies, male Sprague-Dawley rats, 6 to 8 weeks of age, were dosed once daily up to 2 weeks. Routes of administration and vehicle controls were selected per individual compound based on published literature. The duration of studies and sampling times varied according to observed toxicities, although the emphasis was on 6- and 24-h harvest times.

In order to minimize the confounding effect of dose on gene expression, doses were *phenotypically anchored* (standardized) to cytotoxicity *in vitro* and maximum tolerated dose (MTD) *in vivo* as a way of achieving equitoxicity. Both isolated hepatocytes and rats were dosed at two concentrations, designated "high" and "low," with the absolute high and low concentrations different for different compounds. For *in vitro*, the high dose was selected to cause approximately 10% cytotoxicity as determined by alamarBlue assay, which assesses cellular metabolic activity *(12)*. The low dose was selected as the maximum dose at which no statistically significant cytotoxicity is observed by alamarBlue and was estimated by repeated 10-fold serial dilution. For *in vivo*, high dose approximated the MTD, and low dose was an order of magnitude less than MTD. In the context of these studies, the MTD was defined by range-finding studies in Sprague-Dawley rats using in-life observations (body weight, behavior, breathing, posture, etc.), clinical pathology, and histopathology. For many compounds, it is difficult to assess how doses in rats correspond with human exposures and clinical toxicity, as comparative pharmacokinetic/pharmacodynamic data are often unavailable. Essentially, this scheme attempts to choose a high dose where the compound has a clear toxic effect and a low dose where the compound does not. Vehicle controls were collected in parallel for all experiments in hepatocytes and rats. Here, *vehicle* refers to all substances (dimethyl sulfoxide [DMSO], water, etc.) administered as part of delivery of the compound, excluding the compound itself.

Both studies contained a minority of nontoxicants, in part due to the inherent difficulties in drawing a conclusion of *nontoxic*. The criteria used for defining a *nontoxic* was that (a) the compound demonstrates little pharmacologic effect in the liver, and/or (b) the compound has pharmacologic action in liver but demonstrates no hepatotoxicity at physiologically relevant exposures, according to available published literature. The *in vitro* sample consisted of 86 compounds of which 36 were nontoxicants. The *in vivo* sample was slightly larger, containing 111 compounds of which 32 were nontoxicants. The number of compounds shared between the two samples was 52, of which 15 were nontoxicants. Gene expression was measured using the Affymetrix (Santa Clara, CA) RGU34A rat chip, which yielded a total of 8799 expression measures *(13)*. All dose/time combinations were available as repeated observations for both treated and control samples, with an average three replicates for *in vitro* and four for *in vivo*.

4.2. Baseline Sample

Prior to analysis, we chose a high-dose, 24-h design as our "baseline" sample, which meant that these data were chosen as the basis for our final classification rules, with other designs analyzed solely to provide context. The rationale for doing this was twofold. First, use of a baseline sample avoided potential multiple-testing issues that could otherwise have arisen. Second, we believed the high-dose, 24-h design yielded the most powerful strategy for detecting toxicity-induced changes in gene expression *(5)*. Although the addition of earlier time points or low dosage would have increased sample size substantially, their inclusion would have been problematic given that the response variable was literature-based and fixed and thus could not vary according to experimental inputs such as time or dose.

For *in vitro* data, the high-dose, 24-h baseline sample consisted of 277 observations (microarrays) of which 114 corresponded with nontoxicants; that is, $p(\text{nontox}) = .41$. The *in vivo* high-dose, 24-h baseline sample contained 481 observations, with nontoxicants composing 132 of those, and $p(\text{nontox}) = .27$.

5. Analysis Strategy

The raw Affymetrix microarray data (.cel file format) from ToxExpress were prepared for analysis using a multistep procedure. First, the microarray data were processed using RMA (Robust Multichip Average), which performs background correction on individual chips, applies a quantile normalization to force the distribution of probe intensities to be identical across chips, then maps probe intensities to gene expression *(14)*. Second, a small cohort of outliers was removed using a Spearman rank correlation metric. Finally, data were converted to fold change, meaning they were transformed using a ratio statistic in which the numerator is gene expression and the denominator is the mean of a set of time-matched, vehicle control samples.

A tree-based machine learner called stochastic gradient boosting was chosen to analyze the data *(15)*. In brief, the method parameterizes response as an additive expansion in which the basis functions are small trees. The resulting nonlinear optimization is solved using steepest descent *(16)*, and regularization is performed by making smaller than usual steps in the direction of the negative gradient. Boosting retains most of the advantages of tree-based methods with the exception of model simplicity, which it trades in favor of stronger predictive performance. The method allows for both interactions and nonlinear dependency of response on inputs; such effects may be present in genomic pathways. Furthermore, unlike methods such as linear discriminant analysis that have also been applied to toxicogenomics data *(5)*, boosting performs feature selection from the entire set of 8799 genes on the U34A microarray rather than from a much smaller set preprocessed using a dimension reduction technique. This may be important for genes that are highly relevant to mechanisms of hepatotoxicity but that do not contribute large marginal effects on response independent of those of other genes.

The metaparameters used for boosting were 100 trees in the model, shrinkage of 0.1, stochastic subsampling set at 60% of the sample, and tree size limited to two splits. These values were chosen on the basis of our past experience with this algorithm.

Classification rules were fitted to entire baseline samples, with replicate observations treated as independent. Specialized methods exist that can incorporate variance information from replicates for model fitting *(6)*, but they are limited to linear models of response.

For each rule, we estimated its true error *(8,10)* to quantify our degree of belief in the rule *(7)* and to facilitate comparisons between different rules. This component of the analysis was especially important given the hybrid data being used (described earlier) and the nonstandard way in which response was generated. For data with binary response, there is a natural scale for true error, with err = 0 being a perfect rule and err > err_{BCM} likely to indicate a spurious model. Here, BCM is the "Best Constant Model" (sometimes called baseline error) formed by ignoring all inputs (array data) and taking a majority vote solely on observed response *(15)*. If the addition of the array data is not able to improve model performance past err_{BCM}, we conclude no useful information is being extracted from it, and the model—which hypothesizes a functional relationship between inputs and response—is likely spurious *(7)*. Simulation studies in our labs support this notion: when response is randomly permuted with respect to inputs, resulting classifiers have true error approximately equal to that of the BCM (data not shown).

We estimated true error of each classification rule using a bootstrap procedure (cf. **Section 3.2**) modified so sampling was stratified on response level frequencies in the baseline samples and clustered by compound. That is, we defined separate empirical distributions for toxicants and nontoxicants and sampled all observations for a given compound as an indivisible group. We chose a bootstrap estimator of true error due to its superior bias and variance characteristics compared with other designs *(11)*. Although error estimation can be performed ignoring compound-based correlation structure *(6)*, doing so could potentially bias error estimators downwards making classification rules appear more accurate than they actually are.

Bias correction was applied to the bootstrap estimate of true error using the .632+ procedure *(8)*. This method forms a convex combination of resubstitution error, which is biased downwards, and bootstrap error, which is biased upwards. By using these two effects to cancel each other out, the resulting estimator has smaller bias without a significant increase in variance and should have enhanced predictive performance. We denote the .632+ error estimate for the boosting classifier as err_{Boost}.

6. Results

6.1. Classifiers for Toxicity

Separate classification rules were fitted to the *in vitro* and *in vivo* data. Error estimates for *in vitro* classifiers are shown in **Table 1** and are based on a classification cutpoint of 0, the point at which evidence for an observation being toxic or nontoxic is equal (classification cutpoints are described in **Section 6.3**). For the baseline sample ("High dose, 24"), true error for the boosting classifier was estimated as $err_{Boost} = .29$, well below $err_{BCM} = .41$, giving us confidence that the model is real. The boosting classifier comprised 173 unique genes selected from the full complement of 8799 genes. The top 10 genes as determined by a relative importance metric *(15)* are shown in **Fig. 3**. The metric ranks inputs according to their importance in the boosting classifier, with values normalized

Table 1
Estimates of True Error for Various Classifiers of *In Vitro* Data*

Name of design	Genes	Dose	Time(h)	err$_{BCM}$	err$_{Boost}$
High dose, 24	8799	High	24	.41	.29
Candidate genes	156	High	24	.41	.33
Low dose, 24	8799	Low	24	.43	.38
High, 3/6/24	8799	High	3, 6, 24	.43	.39
All	8799	High & low	3, 6, 24	.43	.39

*Estimates of true error for the best constant model and the boosting classifier (respectively, err$_{BCM}$ and err$_{Boost}$) were estimated using the modified bootstrap procedure described in the text.

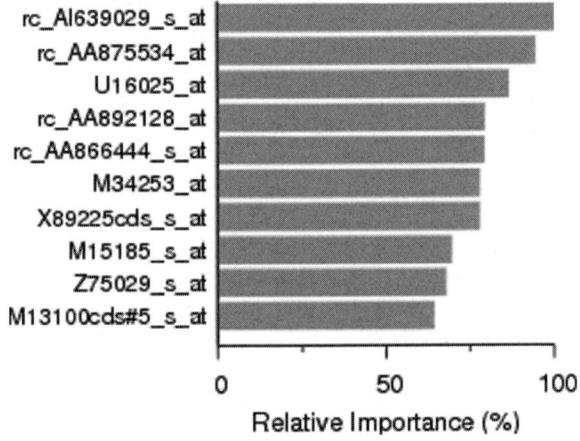

Fig. 3. Top 10 genes in the *in vitro* classifier.

so the most important is 100%. Variables not used in the classifier have relative importance of zero.

For comparison with this agnostic approach, we built a classifier using a candidate gene strategy constrained to 156 genes selected from the literature, in collaboration with Metrigenix U.S., Inc. (Toronto, Ontario, Canada) *(17)*, on the basis of reported involvement in hepatotoxicity. All 156 genes were represented on the RGU34A chip. The resulting boosting classifier used 85 of the 156 genes, and its estimated true error was err$_{Boost}$ = .33, modestly larger than for the agnostic "High dose, 24" approach. Interestingly, there were only three genes in common between the two classifiers.

Additional analyses using all genes were performed incorporating low dose and/or earlier time point data (**Table 1**: "Low dose, 24," "High, 3/6/24," and "All"). In each case, classification error increased substantially, rising close to best constant model values. The results suggest these classifiers are probably spurious and that the genes influencing hepatotoxicity for systems involving low dosage or very early time points are not the same as those for high-dose, 24-h time. Note that values of err_{BCM} differ slightly between designs because not all compounds were run at every dose and time.

Error estimators for *in vivo* classifiers are shown in **Table 2**. The classifier fitted to the baseline sample ("High dose, 24") consisted of 168 unique genes, and its bootstrap error was $err_{Boost} = .22$, modestly smaller than that of the best constant model $err_{BCM} = .27$. The top 10 genes ranked by relative importance are shown in **Figure 4**.

When other designs including the 6-h time point and/or low-dose data were modeled (**Table 2**: "High dose, 6/24," "High/low, 24," and "High/low, 6/24"),

Table 2
Estimates of True Error for Various Classifiers of *In Vivo* Data*

Name of design	Genes	Dose	Time(h)	err_{BCM}	err_{Boost}
High dose, 24	8799	High	24	.27	.22
High dose, 6/24	8799	High	6 & 24	.27	.23
High/low, 24	8799	High & low	24	.26	.24
High/low, 6/24	8799	High & low	6 & 24	.26	.26

*Estimates of true error for the best constant model and the boosting classifier (respectively, err_{BCM} and err_{Boost}) were estimated using the modified bootstrap procedure described in the text.

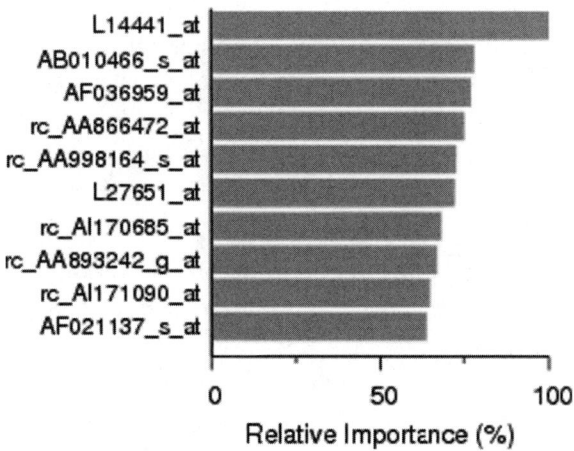

Fig. 4. Top 10 genes in the *in vivo* classifier.

classification error increased slightly, with the worst performance being for the high/low, 6/24 hour design.

Error estimates for all *in vivo* classifiers are relatively close to best constant model values, raising the possibility that some of them are spurious. A confounding factor for interpretation is the lack of balance in the *in vivo* design: almost three quarters of observations are toxicants, thus even the best constant model's simple majority vote of "everything is toxic" gets almost three quarters of predictions correct. In general, the more unbalanced a sample, the harder it is to generate a real (nonspurious) classifier.

To explore this balance issue further, we conducted several additional analyses using the high-dose, 24-h baseline *in vivo* data. In the first, "randomized, high dose, 24," we randomly reassigned response labels of "toxic" and "nontoxic" for compounds while continuing to stratify on the response level frequency of $p(\text{nontox}) = .27$ in the baseline sample. For the second, "balanced, high dose, 24," we randomly removed toxic compounds until the two response levels were equally frequent, that is, $p(\text{nontox}) = p(\text{tox}) = .50$. The procedures were repeated several times, and the average results are shown in **Table 3**.

For the classifier fitted to "randomized, high dose, 24" data, true error was estimated to be .32, slightly worse than BCM levels as expected for a spurious model. This underlines the need for care in interpretation: good classification accuracy ($\sim 70\%$) is not in and of itself an indication of a real model.

The classifier fitted to the balanced training sample had error rates that were clearly smaller than best constant model values (**Table 3**: "Balanced, high dose, 24"), indicating that this classifier is real. Because the balanced data are so closely related to the original baseline *in vivo* sample—being simply a proper subset of baseline data—and because randomization of response resulted in true error clearly higher than that of the baseline classifier, we conclude that the baseline *in vivo* classifier is likely to be real.

Table 3
Comparison of Classifier Performance for Various
***In Vivo* Designs**

Name of design	err_{BCM}	$\text{err}_{\text{Boost}}$
High dose, 24	.27	.22
Randomized, high dose, 24	.27	.32
Balanced, high dose, 24	.50	.35

6.2. Comparisons of In Vitro and In Vivo Classifiers

The *in vitro* and *in vivo* baseline samples share an experimental design (high-dose, 24-h), thus we decided to compare gene lists from their corresponding classifiers—173 genes from the *in vitro* model and 168 from the *in vivo* one. Only five genes were common to both lists, and none of the five was ranked in the top 30 of either classifier as measured by relative importance. If we assume a null hypothesis that the two gene lists were selected independently from the original 8799, the tail probability that they share five members or more is $p = .24$ and we are unable to reject the null. That is, the classifiers for *in vitro* and *in vivo* appear to be essentially different.

The reason for this difference is unclear. Some machine learning methods suffer from variable masking, a trait in which selection of one of a set of highly correlated variables—such as might occur in a genomic pathway—tends to inhibit or *mask* selection of the rest. Supposedly, gradient boosting tends to avoid masking behavior *(15)* by a combination of stochastic subsampling and regularization. In the current study, when analyses were repeated using different random number seeds for stochastic subsampling, there was relatively little change to the classifiers' gene lists, indicating that masking is not a significant factor.

Another potential cause of variation in the classifiers' gene lists are the compounds themselves. Only about half of the compounds were common to both baseline samples, and though this degree of data overlap appears substantial, it is unclear how it translates to overlap between the classifiers.

6.3. Operating Characteristics of the Classifiers

For binary response data, the algorithm used by gradient boosting is related to two-class logistic regression *(15)*. Boosting fits a quantitative response $F(\cdot)$ that takes the form of a log-odds ratio

$$F(x) = \frac{1}{2}\ln\left(\frac{p(y=\text{nontox}|x)}{p(y=\text{tox}|x)}\right) \tag{1}$$

where x is the vector of inputs and y the observed response. By rearranging the terms in Eq. (1), we can construct the following posterior probabilities:

$$p(y=\text{nontox}|x) = 1/\{1+\exp[-2F(x)]\}$$
$$p(y=\text{tox}|x) = 1/\{1+\exp[2F(x)]\} \tag{2}$$

These are estimates for a given observation being toxic or not, conditional upon the gene expression inputs measured for that observation, and based on our classifier $F(\cdot)$.

In order to form a classification rule, $F(\cdot)$ must be dichotomized, and the way this is done determines the operating characteristics of the rule. A rule's operating characteristics can be expressed in one of two ways: conditional upon observations, which yields the familiar false-positive rate (FPR) and false-negative rate (FNR) statistics, or conditional upon predictions, which generates statistics for predictive value positive (PVP) and predictive value negative (PVN). For example, assuming a null hypothesis of "everything is nontoxic," the false-positive rate considers the set of (truly) nontoxic compounds and measures the proportion of toxic predictions within it. In contrast, the predictive value positive considers the set of toxic *predictions* and calculates the proportion of truly toxic compounds within it. For toxicogenomics applications, the PVP/PVN approach is more relevant for revealing the utility of a classification rule because the true toxic potential of a compound is rarely known during hit-to-lead and early lead optimization stages of drug development, at which a screening assay for toxicity would be applied.

Operating characteristics of the baseline *in vitro* classifier are shown as a function of classification cutpoint in **Fig. 5** (*see* Color Plate 5); those for the *in vivo* classifier are given in **Fig. 6** (*see* Color Plate 6). All observations with predicted response $F(\cdot)$ smaller than a given cutpoint are classified as "toxic," whereas observations with $F(\cdot)$ values larger than the cutpoint are "nontoxic."

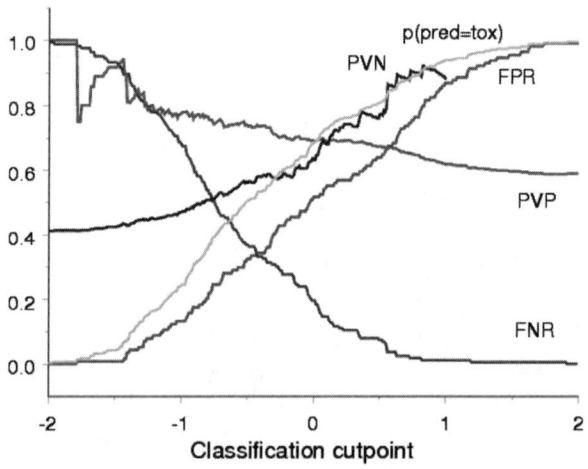

Fig. 5. Operating characteristics of the baseline *in vitro* classifier as a function of classification cutpoint. Replicate observations were treated independently. Statistics shown are FPR, FNR, PVP, and PVN. Also shown is the proportion of predictions called "toxic" (gray). (*see* Color Plate 5).

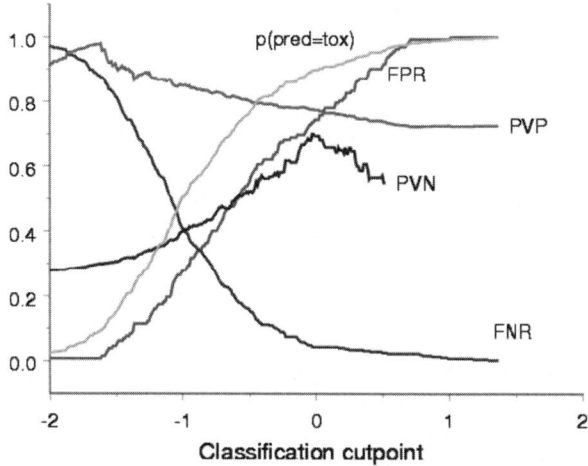

Fig. 6. Operating characteristics of the baseline *in vivo* classifier as a function of classification cutpoint. Statistics shown are FPR, FNR, PVP, and PVN. Also shown is the proportion of predictions called "toxic" (gray). (*see* Color Plate 6).

When the cutpoint is sufficiently large and negative, only those compounds for which evidence for toxicity is overwhelming are classified as toxic. Consequently, FPR is close to 0, FNR is close to 1 (as almost all toxicants fail to meet such a stringent cutpoint), PVP is close to 1 (only the most likely candidates are classified as toxic), and PVN equals the proportion of nontoxicants in the baseline sample. As the cutpoint shifts to the right, the various statistics change as shown in the figures. In order to implement toxicogenomics as a screen, a specific cutpoint could be chosen to optimize a cost function based on factors such as cost of incorrect calls, proportion of hepatotoxic compounds, and so on. The details of such an analysis are beyond the scope of the current discussion.

6.4. Compound-Based Classification

When implementing toxicogenomics as a *compound* screen, we may wish to perform compound-based classification as opposed to mere classification of replicate samples. One way of doing this is to define a compound-based $F(\cdot)$ score, which would then be dichotomized. Given the definition of $F(\cdot)$ in Eq. (1), two compound-based $F(\cdot)$ statistics suggest themselves, namely a mean or median over replicates, with both statistics leveraging the stabilizing effect of replicate observations. We tried both approaches and saw little difference between the two. In general, compound-based classification performed marginally better than its sample-based counterpart (data not shown).

6.5. Compound Similarity Tree

An aim of toxicogenomics beyond mere compound classification is to provide mechanistic insight into toxicologic findings, and such a goal can be explored using clustering algorithms. Although there exist a number of commonly used strategies such as hierarchical clustering that are based on unsupervised learning methods, we chose a different approach. We refitted a single-tree model to sample-level expression data, with the original binary response replaced by the fitted, quantitative response $F(\cdot)$ estimated using boosting *(7)*. The motivation is that $F(\cdot)$ represents a smoothed version of the original binary response and leverages the powerful predictive ability of the boosting classifier, and the use of a single tree allows us to approximate the boosting model in a form conducive to clustering. We expect this approach to be more powerful than unsupervised methods, which by definition ignore response information. Also, this approach is *not* the same as fitting a single tree to the original data, which would ignore knowledge extracted by boosting.

To avoid potential overfitting effects, the response variable used for the single tree was the mean $F(\cdot)$ value taken from our bootstrap out-of-boot samples (an *out-of-boot* sample is the set of all observations not chosen for inclusion in its corresponding bootstrap sample; such samples are used to estimate classification error, as explained above). Tree topology was grown using classification and regression tree *(18)*, as implemented in RPART *(19)*, with size of the final tree determined by cost complexity pruning using 10-fold cross-validation *(18)*. Compound clusters were defined by the terminal nodes of the final tree.

Because the tree was fitted to sample-level data, an additional step was required to generate clusters of compounds. A compound was allocated to a given terminal node of the tree only if at least half of its corresponding sample-based replicates fell into that node. In the case of ties, a compound was deemed to be a member of both the terminal nodes involved. Compounds that failed to fulfill these conditions were excluded.

A compound similarity tree for *in vitro* baseline data is shown in **Fig. 7** (*see* Color Plate 7). The tree topology involves 4 splits using 4 different genes, for a total of 5 clusters defined by the terminal nodes 5, 6, 7, 8, and 9. Nodes 8 and 9 define clusters in which average predicted toxicity is high, node 5 is a cluster with moderate toxicity, node 6 is a cluster in which the mean $F(\cdot)$ value is close to 0 (i.e., neither toxic nor nontoxic), and the cluster defined by node 7 is composed of compounds where evidence for nontoxicity is strong.

The relative importance of the four listed genes in the corresponding boosting classifier was 87%, 79%, 78%, and 20%, respectively. The relatively high values of this statistic give us confidence that this single-tree approximation

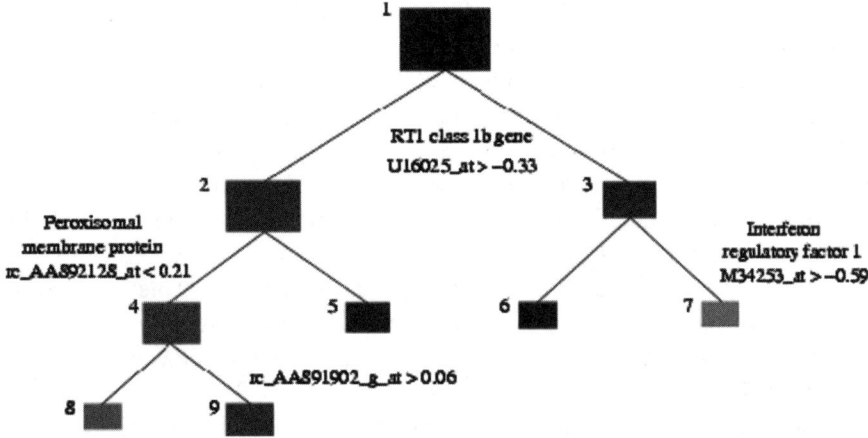

Fig. 7. Similarity tree for *in vitro* compounds. The root node "1" at the top represents the entire "High dose, 24" *in vitro* sample. Beginning at this node, the data are recursively partitioned using fold-change expression values of the indicated genes. Node area is proportional to the number of samples contained in the given node, and color intensity is proportional to the mean $F(\cdot)$ value for the node, with nodes 2,4,5,8 & 9 indicating toxic, nodes 3 & 7 nontoxic, and node 6 a relatively balanced mixture of both. (*see* Color Plate 7).

of the full boosting model is capturing important features of it. Of the 86 compounds comprising the *in vitro* baseline sample, the similarity metric was able to cluster 72 (84%).

7. Discussion

The current report is one of the most comprehensive studies of rat-based toxicogenomics to date, in terms of number of compounds and because it includes both *in vitro* and *in vivo* components. As such, it is of interest to compare our findings to those of other recent studies. Although this is only possible for *in vivo* data, we note that classification error for our baseline *in vivo* classifier was higher than that reported elsewhere *(5,6)*. Several factors may contribute to this finding.

For a start, these other studies only examined one fifth the number of compounds analyzed in the current report, and it is unknown whether their findings generalize to the greater diversity of compounds contained in the current study.

Another confounding factor is the degree of balance (proportion of nontoxicants) in the data, with both cited reports having relatively unbalanced samples (one had only 4% nontoxic compounds). As we showed above, balance can have a major impact upon behavior of binary classifiers. In unbalanced samples, classification error can appear to be small even when response is randomized with respect to inputs, and resulting models must be spurious.

A third factor is the method used to estimate classification error. Although machine learners are a powerful strategy for analyzing genomic data, one of their traits is they *always* produce a prediction rule no matter how tenuous the underlying evidence. This was of particular concern to us given the hybrid experimental design used in this and other studies of toxicogenomics *(5,6)*, in which the response variable—rather than being measured in the same experimental system as the inputs—was merged with inputs post hoc. To quantify our degree of belief in the classifiers constructed, we estimated each classifier's true error using bootstrap sampling combined with .632+ bias correction, we drew bootstrap samples clustered by compound and stratified on response level frequencies in the baseline samples, and we used classification error of the best constant model as a benchmark. Only those classifiers that clearly outperformed best constant model error were considered to be real. Other approaches are feasible *(5,6)* but may not achieve the same accuracy quantifying performance.

In the current study, the experimental design used high-dose and 24-h time of harvest of RNA. When earlier time points and/or low-dose data were included, classification accuracy declined—modestly so for *in vivo* data and markedly so for *in vitro*. The reasons for this performance decrease are unclear and are probably multifactorial; they warrant further study.

Collectively, our findings suggest that classification of the toxic potential of pharmaceutical compounds is possible using either *in vitro* or *in vivo* designs. There were marked differences between the *in vitro* and *in vivo* classifiers with respect to their gene lists, with the degree of overlap consistent with independent selection. However, our analysis included no pathway knowledge of any kind, and classifiers that appear disjoint at the level of individual genes may in fact overlap considerably when viewed with the benefit of pathway information. Nonetheless, if the lack of overlap between the *in vitro* and *in vivo* classifiers persists even when pathway data are taken into account, it would challenge the conventional wisdom that changes in gene expression should transcend biological systems.

Compounds can be rank ordered based on their hepatotoxic potential using either the $F(\cdot)$ values in Eq. (1) or the probabilities of Eq. (2)—the two are equivalent as the probabilities are simple monotonic functions of $F(\cdot)$. Although both the *in vitro* and *in vivo* assays yield robust and actionable information, the *in vitro* assay has several advantages for rapid implementation such as more modest experimental costs, faster throughput, and lower test article consumption. Chemistry and pharmacology activities during hit-to-lead and early lead optimization stages progress relatively quickly, with the data from most *in vitro* and rodent PK assays available within 2 weeks or less, as chemists seek to optimize structure-activity relationships within series members. The *in vitro* toxicogenomic assay has tremendous potential to help compare across competing chemical series, rank order within a series, and provide early mechanistic insight based on a predicted toxicity end point.

The *in vitro* toxicogenomics work described here also raised a number of different technical challenges to be addressed. For hepatocyte cell culture enthusiasts, the choice of substratum, media, and time of culture/treatment are often contentious issues. For the studies in ToxExpress described here, cells obtained from In Vitro Technologies and Cambrex Bio Science were plated on Matrigel Matrix (BD Biosciences, San Jose, CA), cultured in serum-free minimal media, and treated 48 h after isolation. This was an attempt to preserve some of the three-dimensional cellular architecture and clustered morphology, improve viability, and avoid the initial ~24-h postisolation period where the phenotype may be most unstable *(20,21)*. A thorough review of cell culture factors is beyond the scope of this text—readers are directed to Farkas and Tannenbaum *(22)* for an overview of current *in vitro* methods for hepatotoxicity. Over the course of our studies, we explored the use of collagen as substratum and reducing the time of preincubation to 24 h (vs. 48 h) as measures to reduce cost and assay time. Microarray data for several reference compounds within ToxExpress were regenerated and analyzed via the *in vitro* classification rule. Not surprisingly, changes in choice of substratum (collagen instead of Matrigel) and preincubation time (24 h vs. 48 h postisolation) yielded discordant results (data not shown). This does not suggest that other experimental designs will be more or less successful. Rather, this only emphasizes the importance of maintaining consistency in experimental design across training and test data sets.

Finally, we attempted to take a step beyond mere classification and outlined a procedure to cluster compounds based on knowledge extracted during construction of the classification rules. While branding a compound as toxic or not and rank-ordering within a chemical series is undoubtedly of interest, it would be invaluable to discriminate between and predict chemistry- versus mechanism-based toxicity in the liver. Combining the quantitative power of $F(\cdot)$ values with the compound similarity metric may provide valuable guidance for medicinal chemists seeking to improve their molecules. Much work is yet to be done, and we expect this will be the basis for intensive future study.

Typical of any new assay or technology, leveraging toxicogenomics within preclinical toxicology testing paradigms is replete with both technical and philosophical challenges. Despite extensive study of predictive models using reference compound data, it will be difficult to further assess model performance against novel pharmaceutical candidates as only a minority will actually progress to clinical trials. For the models presented in this work, it would certainly be reassuring for a compound identified as toxic in the *in vitro* model to manifest liver pathology in subsequent animal testing, but the ultimate benchmark remains human clinical outcome. As such, the true impact of this technology may not be realized for years to come.

Our belief is that both *in vitro* and *in vivo* approaches are feasible, but *in vitro* has significant potential to evolve into an SAR (structure-activity relationship)-type tool for medicinal chemists, with toxic potential being a novel end point for rank ordering during early lead optimization. This evaluation, however, will require a focused testing approach using numerous examples from the same chemical series contrasted with other series within the same therapeutic target program.

Another obvious area for future exploration is the platform used to capture gene expression changes. Measuring gene expression by hybridization arrays is semiquantitative, reasonably costly, generally low-throughput, and often regarded as being "noisy" compared with the increased specificity and sensitivity of RT-PCR. It is interesting to speculate whether the 173 *in vitro* model genes described above would yield comparable or improved classification accuracy when carried forward to a quantitative PCR platform or emerging technologies based on chromatin immunoprecipitation or branched DNA. Nonetheless, there is clearly a need for new methods and capabilities within preclinical toxicology, and toxicogenomics has demonstrated tremendous potential to improve compound pipelines and mitigate toxicologic liabilities earlier during development.

Acknowledgments

The authors thank our colleagues at Millennium Pharmaceuticals, Inc.—Arek Raczynski, Carl Alden, and Scott Coleman—for their scientific input and thoughtful review of this work; and Victor Farutin for his expert help implementing stochastic gradient boosting. We also thank Gene Logic Inc. for supplying the *in vivo* and *in vitro* gene expression data. Permission to reproduce material was obtained from Future Medicine for the inclusion of parts of the following papers from Pharmacogenomics:

- Barros, S. (2005) The importance of applying toxicogenomics to increase the efficiency of drug discovery. *Pharmacogenomics* 6(*6*), 547–550.
- Martin, R., Rose, D., Yu, K., and Barros, S. (2006) Toxicogenomics strategies for predicting drug toxicity. *Pharmacogenomics* 7(*7*), 1003–1016.
- Martin, R. and Yu, K. (2006) Assessing performance of prediction rules in machine learning. *Pharmacogenomics* 7(*4*), 543–550.

References

1. Kola, I. and Landis, J. (2004) Can the pharmaceutical industry reduce attrition rates? *Nat. Rev. Drug Discov.* **3**, 711–715.
2. DiMasi, J., Hansen, R., and Grabowski, H. (2003) The price of innovation: new estimates of drug development costs. *J. Health Econ.* **22**, 151–185.
3. Dambach, D.M., Andrews, B.A., and Moulin, F. (2005) New technologies and screening strategies for hepatotoxicity: use of *in vitro* models. *Toxicol. Pathol.* **33**, 17–26.
4. Kaplowitz, N. (2005) Idiosyncratic drug hepatotoxicity. *Nat. Rev. Drug Discov.* **4**, 489–499.
5. Raghavan, N., Amaratunga, D., Nie, A., and McMillian, M. (2004) Class prediction in toxicogenomics. *J. Biopharm. Stat.* **15**, 327–341.
6. Natsoulis, G., Ghaoui, L., Lanckriet, G., Tolley, A., Leroy, F., Dunlea, S., et al. (2005) Classification of a large microarray data set: algorithm comparison and analysis of drug signatures. *Genome Res.* **15**, 724–736.
7. Breiman, L. (2001) Statistical modeling: the two cultures. *Stat. Sci.* **16,** 199–231.
8. Efron, B. and Tibshirani, R. (1997) Improvements on cross-validation: the .632+ bootstrap method. *JASA* **92**, 548–560.
9. Hastie, T., Tibshirani, R., and Friedman, J. (2001) *The Elements of Statistical Learning*. Springer-Verlag, New York.
10. Efron, B. (1983): Estimating the error rate of a prediction rule. *JASA* **78**, 316–331.
11. Martin, R. and Yu, K. (2006) Assessing performance of prediction rules in maching learning. *Pharmacogenomics* **7(4)**, 543–550.
12. Fields, R.D. and Lancaster, M.V. (1993) Dual-attribute continuous monitoring of cells proliferation/cytotoxicity. *Am. Biotechnol. Lab.* **11**, 48–50.
13. Affymetrix RGU34A microarray. Available at www.affymetrix.com/analysis/index.affx.
14. Irizarry, R., Hobbs, B., Colin, F., Beazer-Barclay, Y., Antonellis, K., Scherf, U., and Speed, T. (2003) Exploration, normalization, and summaries of high density oligonucleotide array probe level data. *Biostatistics* **2**, 249–264.
15. Friedman, J. (1996). Greedy function approximation: a gradient boosting machine. *Ann. Stat.* **29,** 1189–1232.
16. Luenberger, D. (2003) *Linear and Nonlinear Programming*, 2nd ed. Springer, Berlin.
17. Metrigenix. Corporate home page. Available at www.metrigenix.com.
18. Brieman, L., Friedman, J., Olshen, R., and Stone, C. (1998) *Classification and Regression Tree*. Chapman and Hall/CRC, Boca Raton, FL.
19. Therneau, T. and Atkinson, E. (1997) An introduction to recursive partitioning using the rpart routines. *Mayo Clinic Technical Report*.
20. Baker, T., Carfagna, M., Gao, H., Dow, E., Li, Q., Searfoss, G., and Ryan, T. (2001) Temporal gene expression analysis of monolayer cultured rat hepatocytes. *Chem. Res. Toxicol.* **14**, 1218–1231.

21. Morghe, P., Berthiaume, F., Ezzell, R., Toner, M., Tompkins, R., and Yarmush, M. (1996) Culture matrix configuration and composition in the maintenance of hepatocyte polarity and function. *Biomaterials* **17**, 373–385.
22. Farkas, D. and Tannenbaum, S. (2005) In vitro methods to study chemically-induced hepatotoxicity: a literature review. *Curr. Drug Metab.* **6**, 111–125.

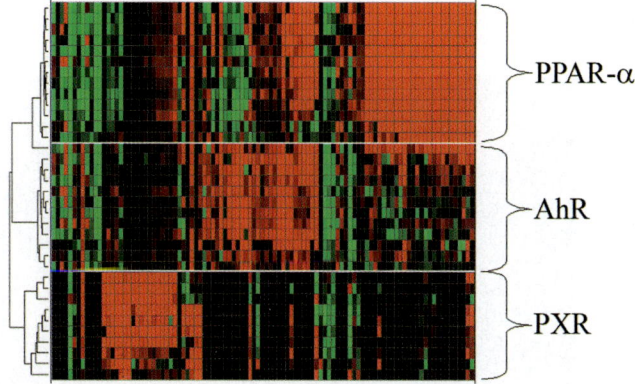

Color Plate 1. Identification of genes regulated in the liver of rats after xenobiotic activation of the nuclear receptors PPAR-α, aromatic hydrocarbon receptor (AhR), or pregnane X receptor (PXR). The heatmap shows the genes significantly regulated in liver by several prototypical inducers of the three nuclear receptors. Genes shown in red are upregulated relative to vehicle-treated control, genes shown in green are downregulated relative to vechile-treated control, and genes shown in black are unchanged. These data were extracted from the Iconix DrugMatrix database. (Chapter 2, Fig. 1).

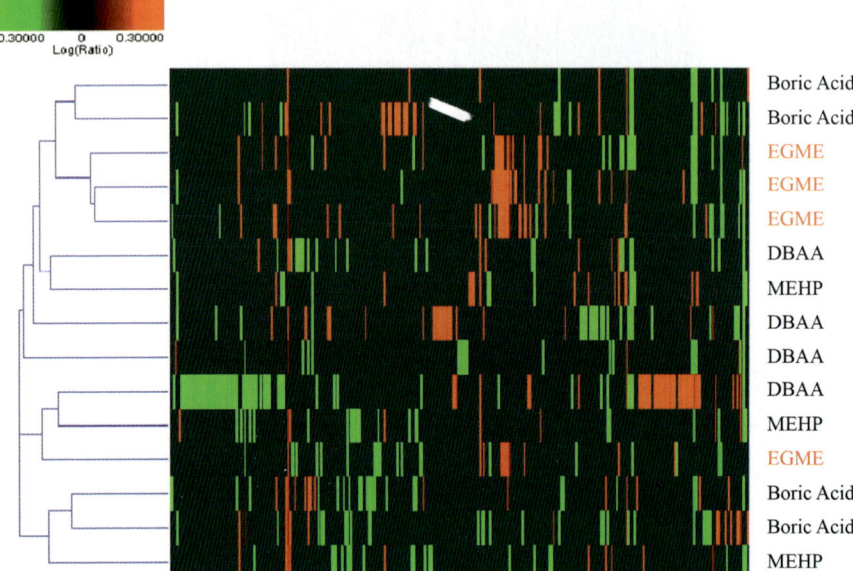

Color Plate 2. Hierarchical clustering of gene expression profiles of the testes of male Sprague-Dawley rats treated with a single dose of various testicular toxicants and sacrificed 24 h after treatment. Included in the heatmap are genes that were upregulated or downregulated by a factor of ≥1.5 with a *p* value less than .01. Green indicates downregulation, and red indicates up-regulation. In general, the testicular toxicants induced limited numbers of consistent gene expression changes after 1 day of treatment, suggesting that the treatment period was too short to generate a list of differentially expressed genes that could be used for mechanistic understanding. EGME, ethylene glycol monoethyl ether; DBAA, dibromoacetic acid; MEHP, mono-(2-ethylhexyl) phthalate. (Chapter 2, Fig. 2).

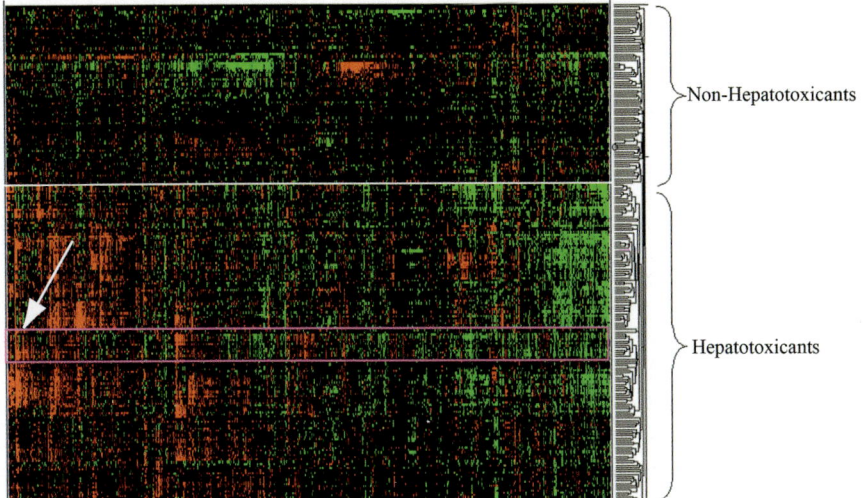

Color Plate 3. Heatmap of gene expression profiles from the liver of rats treated with Cpd-001 (arrow) and a wide variety of reference compounds including nonhepato-toxicants and hepatotoxicants. The expression profiles induced by Cpd-001 are highly similar to and cluster with the expression profiles from the reference hepatotoxicants in the Iconix DrugMatrix database. Green indicates downregulation, and red indicates upregulation. (Chapter 2, Fig. 4).

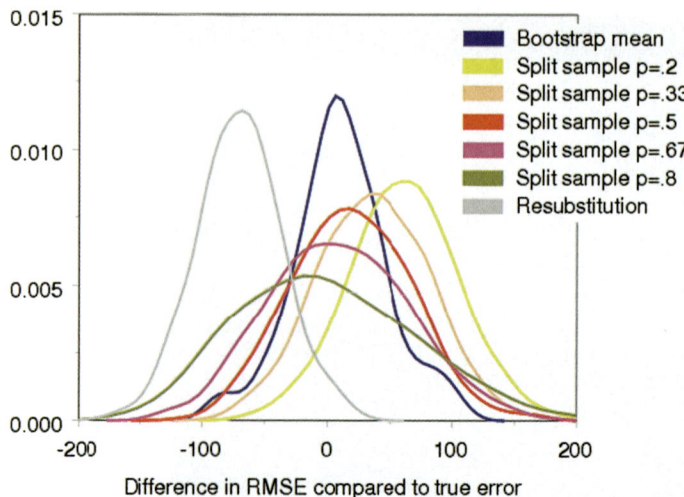

Color Plate 4. Distributions for error estimators based on proteasome data. The graphs plot the difference between the estimated error and the true error. Parameter p is the proportion of a baseline sample used for training during error estimation. (Chapter 5, Fig. 2).

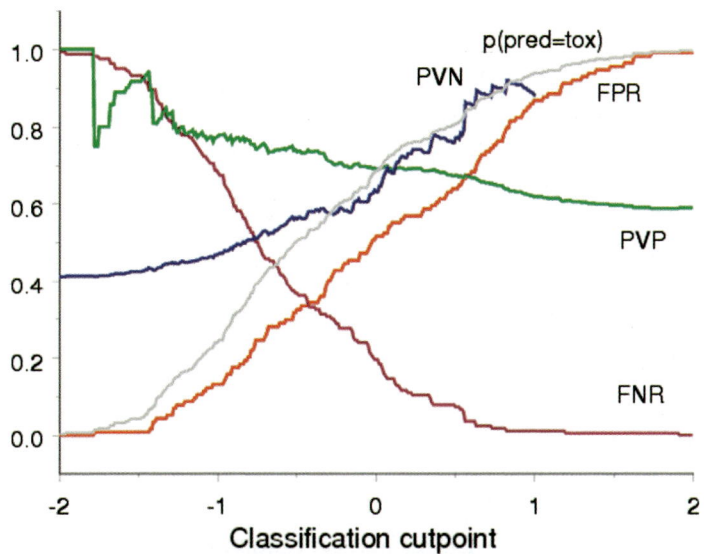

Color Plate 5. Operating characteristics of the baseline *in vitro* classifier as a function of classification cutpoint. Replicate observations were treated independently. Statistics shown are FPR (red), FNR (magenta), PVP (green), and PVN (blue). Also shown is the proportion of predictions called "toxic" (gray). (Chapter 5, Fig. 5).

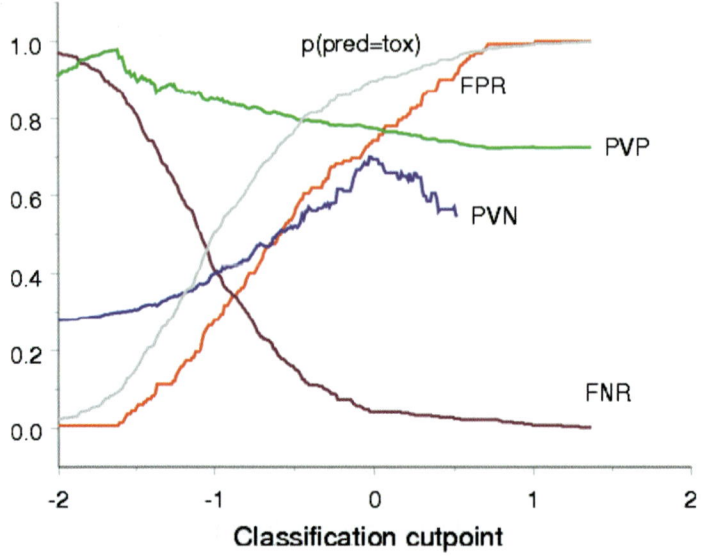

Color Plate 6. Operating characteristics of the baseline *in vivo* classifier as a function of classification cutpoint. Statistics shown are FPR (red), FNR (magenta), PVP (green), and PVN (blue). Also shown is the proportion of predictions called "toxic" (gray). (Chapter 5, Fig. 6).

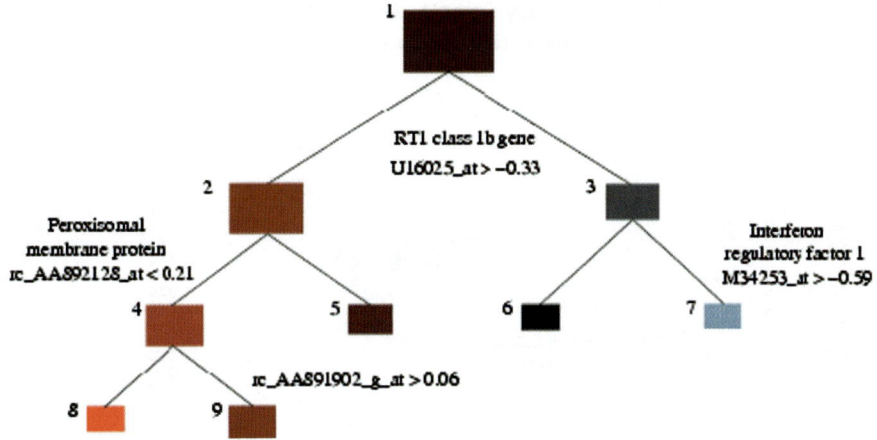

Color Plate 7. Similarity tree for *in vitro* compounds. The root node "1" at the top represents the entire "High dose, 24" *in vitro* sample. Beginning at this node, the data are recursively partitioned using fold-change expression values of the indicated genes. Node area is proportional to the number of samples contained in the given node, and color intensity is proportional to the mean $F(\cdot)$ value for the node, with red indicating toxic, blue nontoxic, and black a relatively balanced mixture of both. (Chapter 5, Fig. 7).

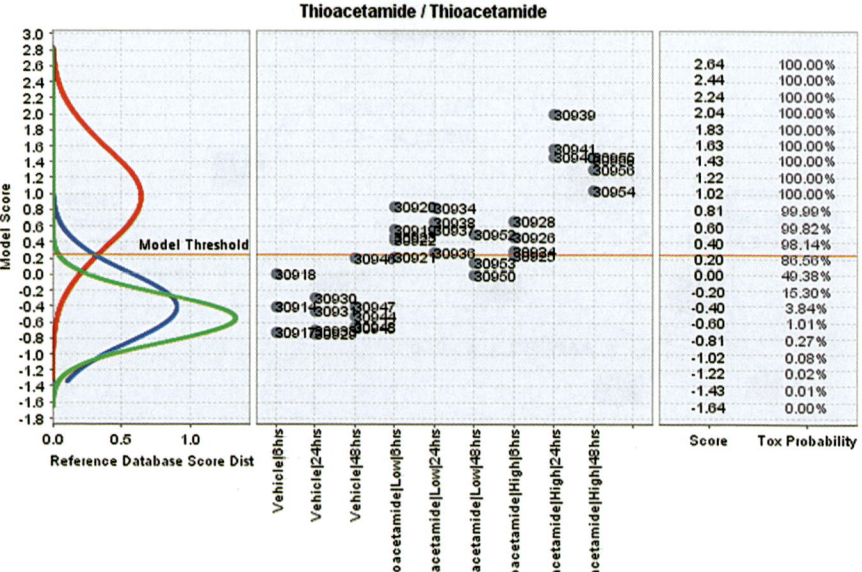

Color Plate 8. Model scores for two doses of thioacetamide or vehicle-treated samples at 6-, 24-, and 48-h exposures. The plot from Gene Logics ToxExpress System database is provided as an example of a diagnostic visual of toxic response based on a model system. A full description of the plot is provided in the text in Section 2.2.6.4. In short, the leftmost portion of the plot shows the behavior of reference database known toxicant-treated samples (red), known non–toxicant-treated samples (blue), and vehicle-treated samples (green) as a function of the output model score. The model in this case is assessing general liver toxicity of any type. The rightmost portion of the plot shows the probability of toxicity as calculated based on the overlapping distributions seen in the leftmost portion. The middle portion of the plot shows the experimental samples plotted relative to the model score. Each treatment group is divided by the x axis and individual samples within each treatment group are labeled. The horizontal line is the binary threshold above which a response should be deemed "toxic." The continuous nature of the modeling output is evident and provides critical information such as intragroup versus intergroup variability, dose and time trends, proximity of response to the binary threshold, and relative confidence based on reference database behavior. All of this information is critical to providing correct interpretation for a modeling result. (Chapter 6, Fig. 4).

6

In Vivo Predictive Toxicogenomics

Mark W. Porter

Summary

Reference databases consisting of large sample numbers and high-dimensional microarray data are now available for the investigation of adverse events in animal model systems such as the rat. This large volume of data, accompanied by appropriate study designs, compound and dose selection procedure, and minimization of technical and biological confounding effects, can yield successful predictive models for a variety of hypotheses. The process of training, validating, and implementing predictive models is cyclical and complex. This chapter highlights individual decisions that need to be made before, during, and after a model or set of models has been trained, with an emphasis on proper statistical methods and suitable interpretation of the results.

Key Words: classification; *in vivo*; microarray; modeling; prediction; toxicity; toxicogenomics.

1. Introduction

The pharmaceutical industry is currently challenged with using new technologies to advance safe, efficacious drugs through the development process in a faster and cheaper fashion than what has been realized based on traditional methods. One candidate technology for long-term incorporation within drug development is the application of genome-wide transcription measures from animals treated with compounds. Gene expression–based platforms offer the potential both to predict adverse events likely to occur in animals and/or humans and to capture the underlying mechanistic events associated with toxicity in a variety of target organs and cell types. This potential has been fully enabled recently through the development of high-throughput

From: *Methods in Molecular Biology, vol. 460: Essential Concepts in Toxicogenomics*
Edited by: D. L. Mendrick and W. B. Mattes © Humana Press, Totowa, NJ

microarrays containing tens of thousands of genes and with the advent of large reference databases composed of transcriptional profiles of hundreds of compounds and many thousands of samples.

This chapter focuses on predictive modeling as a tool for the identification and characterization of the toxic response at the molecular level using microarray-based gene regulation events as the primary data type. The emphasis of the review is on the ability of models to predict toxic events using an *in vivo* study design for a variety of target organs. Although mechanistic and predictive toxicogenomics are intimately related, they are fairly distinct with respect to the types of questions answered, the types of tools used, and the ways in which they can be implemented. Mechanistic events should never be ignored while applying predictive models to a developing compound. However, for the informational purposes of this review, methods of predictive modeling will be highlighted without regard to specifics concerning mechanisms of toxicity. This allows for a general treatment of the subject as it applies to a wide variety of organ systems and situations.

1.1. Feasibility and Utility of Toxicogenomics Predictive Modeling

The combination of microarrays, toxicology studies, and predictive modeling provide far-reaching advantages over classic toxicity panels currently in use. Microarrays offer tens of thousands of variables (i.e., genes or fragments of genes) to be used in the predictive set, as opposed to a few dozen classic indicators. It is important to view as many molecular events as possible on a genome-wide basis in order to build a robust model given the large number of potential toxicity mechanisms within an organ system, the differences in dynamics of drug uptake and launch of specific mechanisms, and the many confounding factors that are known to affect development of toxicity. The selection of genes that completely characterize both toxicity-related events and events unrelated to the toxic response across time necessitates a look at the full genome.

The predictive quality of toxicogenomics data stems from observations that, although gene expression profiles are different for each compound when administered in toxic doses, compounds inducing similar mechanisms of toxicity yield similar global gene expression profiles *(1–3)*. Specifically, groups of genes that successfully discriminate between different toxicity end points and nontoxic responses have been reported *(4,5)*. Although this homogeneity in gene expression response is commonly observed, it cannot be immediately generalized to all common toxicity end points between compounds and, as one would expect, each hypothesis must be explored independently to fully determine whether it is reasonable to implement a predictive model for a certain grouping of compounds. This exercise can be facilitated by unsupervised clustering methods in order to capture which types of compounds are most closely correlated on a global gene expression basis and are best suited for predictive modeling.

The benefit of predictive modeling in toxicogenomics lies in its ability to estimate a probability that a certain compound or series of compounds will cause toxicity within one or more target organs and potentially across species. This probability is usually based on some continuous metric (e.g., a model score) that defines a range and distribution of those scores on compounds of known toxicity potential. For an incoming test compound, the greater the deviation of the score in one direction relative to the separating plane, the higher the toxicity potential for the compound. Scores in the opposite direction indicate lower toxicity potential. This simple paradigm can be extended by not only assaying for general toxicity but also building models for types of toxicity pathologies, changes in clinical profiles, and pathway- and mechanism-specific processes. In fact, most predictive modeling systems contain more than one classifier to yield entire sets of predictive profiles that calculate probabilities for a diverse set of toxicity indications.

2. Overview of Predictive Modeling
2.1. The Predictive Modeling Process

Predictive modeling can be thought of as a general process that can be applied to any number of disciplines and data types. A general outline of the process is presented in **Fig. 1** with six major steps contributing to its successful implementation in a progressive fashion:

1. *Hypothesis and goal development*: Defines overall expectations for the predictive model or set of models. Factors such as types of toxicity, organ systems, species-specificity, confounding factors to be avoided, and so forth, all need to be considered. One very important factor to consider is the type of study designs and protocols that will be used in the resulting assay in order to ensure that all downstream steps are geared toward the support of those study designs and protocols.
2. *Data collection*: Defines the type of data required to build the models defined in **step 1** and includes a determination on what data is currently available, what data needs to be created in the lab, and what homogeneity of data is desired (e.g., single or multiple strains, genders, routes of administration, microarray platforms, sacrifice times, dosing levels and regimens, etc.).
3. *Data parsing*: Defines the way in which data will be used to train the individual models relative to the primary goal for each model. This step includes a determination of which compounds will be used and how they will be used as well as how

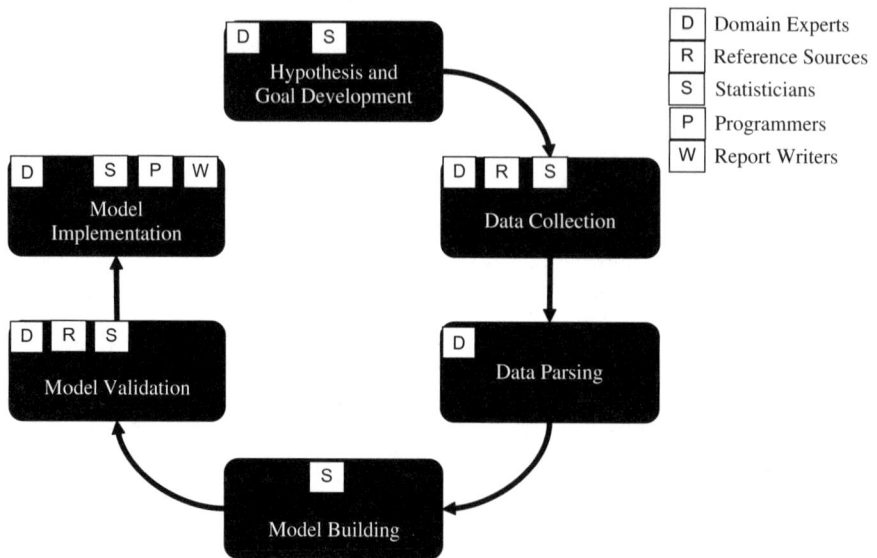

Fig. 1. The Generic Predictive Modeling Process. This schematic attempts to show the entire process of predictive modeling and shows where expertise is necessary within each step. Although the process is shown as linear, in reality it is highly convoluted given that lessons learned from any one area can influence changes to any of the others. Letter coding indicates the contribution of various individuals for a toxicogenomics predictive model at each of the steps. The "Training Set Database" normally comprises a rather large number of people who run animal experiments, process microarrays, and manage data.

   individual treatment groups (i.e., doses and time points) within those compound studies will be used.

4. *Statistical methodology*: Defines the types of microarray data (e.g., normalization method, relative fold change vs. absolute signals, etc.) to be used as well as the gene selection method and classification scheme(s) to be used.

5. *Model validation*: Defines the strategy by which the models are validated. Strategies include both cross-validation techniques such as bootstrapping-like methods, as well as external validation using compounds that were not a part of the model build. The researcher should know in advance what is acceptable and unacceptable in terms of sensitivity and specificity measures.

6. *Model implementation*: Defines the process by which individual researchers will access the models, upload data, view results, provide and record interpretations, and archive all deliverables. Also defines the types and format of results to be presented (e.g., tables, visuals). In addition to raw model results, interpretive aids such as intuitive visuals and tables help to alleviate concerns with the predictive modeling "black box."

The process is inherently cyclical. In addition to a predictive model's inferential properties, it also allows for further research of various hypotheses and methods by tweaking or perhaps even completely changing hypotheses or goals, the way in which data is collected to support prediction, the parsing strategy, and how the model is validated and implemented. Based on the results of previous modeling exercises, further improvements are made at any one of these steps. In fact, it is highly unlikely that the first iteration of this process for a particular data set or hypothesis will yield an acceptable model. It is usually rather obvious that a few changes to any of the first four steps of the six will yield a more reliable model.

The modeling process within a toxicogenomics environment is highly collaborative in nature, as it is dependent on a wide variety of expertise in such areas as toxicology within the organ system(s) being studied, general molecular toxicology, biostatistics, and bioinformatics. Depending on how scientists are to access the model, one may additionally need to deploy programmers and report writers during the implementation phase of the process. Although all of these functions can be performed by a single individual, this individual would require at least rudimentary knowledge for each of these areas. An example of the types of individuals and where they best contribute to the six steps is also shown in **Fig. 1**.

2.2. The Predictive Modeling Process for In Vivo Toxicogenomics

This section will discuss each of the six steps defined in the previous section as they relate to *in vivo* toxicogenomics. The overall goal of predictive toxicogenomics can be stated easily enough as the prediction of whether a compound will be toxic to a specific organ system and, if so, the prediction of the type of toxicity induced. Despite this well-understood concept, predictive toxicogenomics is a fairly new field that has not yet stabilized enough to produce industry-wide common ground as to what is biologically feasible, how to measure the success of the technology, and where or how the technology can be incorporated into existing drug development pipelines. All of these ambiguities result in a wide variety of scientific opinion on study design (both for model-building and assaying), species specificity, overall value, and regulatory stance.

2.2.1. Hypothesis and Goal Development

There is a large number of possible toxicity events that can be modeled based on gene expression information within many different organ systems. Two factors should drive which of these to explore: science-driven hypothesis testing and value-driven goals. The first factor deals with what is biologically possible given the data. Not all gene expression events at the time and dose points to be measured may yield successful prediction for observable toxicity end points. The second factor deals with a more business-focused decision concerning which predictive models should be investigated to provide the most value. End points that have been unsuccessfully predicted using classic safety measures will benefit the most from investigation of alternate measures such as gene expression. In reality, a combination of both strategies works best to yield highly accurate models that supplement existing traditional assays.

2.2.1.1. Organ to Be Assayed

The choice of organ system to assess for toxicity can be made on a variety of factors related to the overall value and feasibility of using gene expression data versus current, traditional techniques. A few of the more common systems to be tested recently within a predictive environment are liver, heart, and kidney.

Feasibility concerns include the homogeneity of the cell populations resident within the organ being modeled. Kidney and heart are more complex than the liver in terms of the distribution of cell types composing them. Although techniques are available to isolate specific cell populations to provide a more homogeneous response than a gross section, they add complexity into the processing steps for the overall assay and have been shown to add process variability into the data *(6)*. Additionally, if these techniques are used as part of a model training set, the test sets will also require this extra step of sample processing, adding more to the cost of the resulting assay. Another feasibility concern is that of separating efficacy responses from a toxicity response. Toxicity can occur within the same pharmacological target organ, as evidenced by the cardiotoxic effects of numerous catecholamines *(7)*. Once a modeling system for a particular organ type is considered, a review of the overall degree of efficacious response versus toxic response comprised by the training compound set is necessary.

A few key types of toxicity events that have been investigated in the liver through predictive modeling and have shown considerable promise in their success include necrosis, steatosis, cholestasis, enzyme induction, peroxisome proliferation, genotoxicity, and more *(8–10)*. Although not much literature exists for kidney and heart toxicogenomics predictive modeling, a variety of end points have been modeled with various levels of validation as part of unpublished commercial services. These include modeling cardiac adrenergic agonist behavior, myocardial necrosis, and arrhythmia, as well as renal Blood Urea Nitrogen (BUN) levels and other kidney toxicity-associated biomarkers.

Surrogate markers of toxicity in blood are currently under general investigation in several laboratories, though none have published findings to date on whole-genome predictive modeling. Monitoring of transcript levels in peripheral blood to identify distal organ toxicity within animal models is the first step in a more extensive validation in order to implement gene expression technology within the clinical environment. Transcript biomarkers in the blood may yield consistent expression profiles of specific organ toxicities and perhaps subtypes of those general toxicities. These transcript biomarkers may either precede or be coincident with classic toxicity measures within the target organ. Two important technological limitations in blood profiling have recently been addressed to increase the viability of whole peripheral blood transcript assays. PAXgene Blood RNA System (PreAnalytiX) reduces process-induced activation of white blood cells and offers a reliable long-term storage system *(11)*. New globin reduction protocols remove the overwhelming amount of globin mRNA from the whole-blood population, which increases both sensitivity and reliability of whole genome blood assessments. With these technical improvements, the data collection portion of the process is improved, but it remains to be proven that blood transcripts are reliable predictors of various toxicity responses.

2.2.1.2. BENCHMARKS WITH RESPECT TO DOSE AND TIME

mRNA transcripts can be regulated preceding classic end points such as clinical chemistry measures or histopathology findings due to mechanisms that may be causative of, or just generally related to, toxicity. This allows for prediction of toxicity with respect to time course using gene expression assays. In addition, gene expression–based markers may be more sensitive predictors of toxicity at lower doses than what is necessary to elicit toxicity, allowing for prediction of toxicity with respect to dose levels. In both cases, the value is clear for toxicogenomics-based assays to potentially enable shorter treatment times in animal studies, requiring less animal husbandry and compound per study, and to enable lower doses, further reducing the amount of compound required. The researcher should decide what type of classic benchmark(s) will be used tboth to build a training set of treatments and to assay test compounds. This assessment will greatly affect the utility of the model once it is built. For instance, gene selection will be different for markers that are correlative with classic end points, as opposed to markers that are predictive of end points.

There are several published examples of marker sets whose regulation precedes end-point pathology or other measure. The liver carcinogenicity assays currently employed within preclinical animal studies normally require years of dosing with each compound under investigation. Recent reports using known long-term carcinogenicity agents in short-term studies (i.e., less than 1 week of dosing) demonstrate that transcripts and pathways associated with carcinogenicity are regulated at very early stages of treatment *(4,12)* and may even distinguish between genotoxic and nongenotoxic types of long-term carcinogenicity *(13)*. Toxicity-mediated gene expression regulation has also been observed to precede a variety of other pathologies in multiple organs such as liver and kidney using well-known toxicants *(3,14)*. Though not as valuable as carcinogenicity testing in terms of time savings for a full classic study, it should be noted that many of these markers for more acute pathologies were regulated within a 1-week period. This similarity in study design for the capture of multiple toxicity responses in a short *in vivo* study shows promise for a common assay to predict multiple phenotype development of short, moderate, and long terms.

For all of the aforementioned studies, it is important to note that a great deal of validation on these potential markers is yet to be done to ensure a true predictive response across a large number of toxicants.

2.2.1.3. Species Specificity

Inability to correlate safety findings from one species to the next remains one of the most significant challenges in toxicity testing. The ability to measure molecular events, rather than classic end points, for the launch of specific mechanisms within an animal model would aid in the identification of potential problems before clinical trials are initiated. If predictive assays are implemented at time points early enough to measure gene expression changes common to multiple species before the modification of the response with species-specific metabolites, for instance, one can capture a more transferable cross-species toxic response. Although cross-species inferencing has not been shown in peer-reviewed publications, it has been claimed by several groups that models robustly predict across species. Although some models appear to be robust in predicting human-specific events, preliminary investigations of species-specific gene expression–based mechanisms have so far turned up nothing to provide explanations of why this occurs. As an added concern, the definition of a "human-specific" toxicant is not clear and leads to discussions concerning incidence, severity, and dosing relationships between efficacy and toxicity for each of the supposedly human-specific compounds. An increase in the amount of literature showing success at predicting toxicity specific to one species and demonstration of mechanistic understanding of the phenomenon will aid greatly in justifying incorporation of gene expression data into the drug development pipeline for such purposes. To date, these fundamentals have not been proved, partially due to disagreement over the definition of species specificity using classic end points.

2.2.1.4. Confounding Effects

With the multiple dependencies of individual gene and pathway regulation on a variety of biological processes, many factors need to be considered that could be mistaken for a toxic response when using gene expression measures. These factors include, but are certainly not limited to, animal strain and breed, gender, pharmacological effects, toxicity effects external to the organ being measured for toxicity, circadian and diurnal variation, and fasting. The list of potential confounding effects is too large to be able to include all factors within a modeling system. However, it is possible to estimate the prevalence of the largest effects based on previous literature and reference databases where specific studies have been conducted to assess these factors. Alternately, incorporating additional variables into the data collection and model-building processes can control these effects. In general, if the model has been appropriately trained using these confounding factors, the model will be more insensitive to these confounding effects than if it had not.

The ultimate goal of the eventual model will be to assay true test compounds on an ongoing basis as part of the primary drug development process. The diversity of the samples and study designs to be assayed as part of this process will determine the number of confounding effects that need to be assessed. For commercial predictive service providers such as Gene Logic Inc. (Gaithersburg, MD, USA) and Iconix Pharmaceuticals Inc. (Mountain View, CA, USA), which both offer predictive services for models built on compound data from a single reference database, samples may be tested from animals generated and microarrays processed outside of its database parameters. Therefore, the sensitivity of the models to many types of site-specific and process-based confounding effects must be rigorously addressed. In other cases, the models may need to support more focused study designs where all data is extracted from a single source and is only applied to specific and highly controlled study designs such as single strain, gender, route of administration, sacrifice time-of-day, and so forth. For these cases, the assessment of confounding effects need only address those variables that remain uncontrolled. In an optimal, well-controlled drug development process where the model is applicable to a very specific portion of the pipeline, one may achieve a system where only day-to-day variability of the in-life and microarray processing portions of the experiment require monitoring, as all other processing site-based confounding effects are controlled within the experimental design.

2.2.1.5. INTERPRETABILITY AND ACCURACY

Within any genomics-based predictive model, genes are normally used as predictor variables to classify samples by treatment or disease state. Biologists are wary of algorithms that, although highly accurate, are based on large collections of genes, some of which may already be known to contribute to the biology, but most of which are unknown. The claim is rightly made that reliable predictors can increase understanding of biological systems at the molecular level solely based on their high degree of concordance to the end point. However, this is of limited comfort to those researchers who have received an unexpected result from the model and simply want to know why a particular outcome has occurred for their sample(s). The explanation may not be inherent in the known functions of the markers responsible for the prediction.

From a purely statistical view, a microarray contains tens of thousands of variables that should each be used to yield the most accurate assay. As arrays are expensive to run, it theoretically does not make sense to throw away good data. Usage of a genome-wide system based on a large number of compound treatments will yield the most accurate models inherent to the data. Prefiltering the potential list of predictors based on severely limited researcher knowledge will only reduce that accuracy. Gene selection techniques and characteristics of optimal marker genes will be described in **Section 2.2.4**.

On the other hand, from a feasibility perspective, highly multidimensional black boxes are not easily interpretable, especially when the result is given as a single score and probability of class membership. The behavior of the underlying genes is important in explaining how the result was achieved. Not only does this approach call for a reduction in the overall number of predictor variables such that their individual contributions to the resulting call are more easily observed, but it also calls for a reduction of the predictors to those that are understood by the industry in general. These "mechanistic signatures" are normally composed of a few dozen genes, at least a sizeable portion of which are already known to contribute to the toxicity or disease state being assayed. Although this approach rarely yields the most accurate model from the microarray, it may gain acceptance more quickly by more traditional toxicologists and is transferable to smaller platforms such as multiplex PCR. Because these mechanistic signatures do not yield optimal models, they are more appropriately assessed by a more generic *change score* that defines the degree to which regulation occurs within the signature, rather than a predictive modeling assay using a classifier and resulting in probabilities of toxicity. This change score can be used to rank the mechanisms that are most important for further review but does not imply predictive qualities.

The problem of interpretability and accuracy can be addressed in multiple ways, two of which are outlined here. First, the trade-off between gene set size and model accuracy can be incorporated into the model-building algorithms themselves so that a compromise between the two can be reached *(15)*. In this method, a controlling parameter is a part of the classification algorithm and looks for optimal success rate in relation to the gene size. The researcher(s) can then decide what drops in accuracy are acceptable given reduced gene set size in order to implement a final model. The disadvantage to this method is that the implemented predictive model will not provide the most accurate results possible, and genes contributing to the actual toxicity when generalized to the full universe of compounds are likely not included as predictors. Second, the most accurate model can be built using as many genes as necessary, but interpretive aids can be implemented so that a researcher can determine which markers contributed most to the final prediction. This method allows for a model that can be better generalized to the universe of compounds, as all good predictor genes are included, and allows for a separate filtering of the genes that were most responsible for a prediction. The disadvantage to this method is that statisticians building the model must take the extra steps necessary to construct the additional filtering tools, visuals, and so forth.

An example of the latter is shown in **Fig. 2**. In this situation, marker genes are used to predict a sample as generally toxic to the liver at an approximate 90% accuracy range based on both cross-validation and validation with external samples. A statistic called the *gene contribution* has been calculated such that the weightings of each marker gene from the model itself are applied to the incoming sample's regulation of that gene. The resulting contribution is high if both the gene's weight is high and the degree of regulation in the incoming sample is high. From the distribution of gene contributions shown in the figure, most genes are weighted close to zero such that they do not contribute to the predictive call (i.e., they are either not highly weighted in the model or are not highly regulated by the treatment). However, some genes contribute greatly to the predictive call and can be considered the markers that most contributed to the overall prediction (i.e., they are both highly weighted in the model and are highly regulated by the treatment). A review of this list of markers may indicate the mechanism by which the toxicity will be achieved. At this level, prediction and mechanism can both be assessed, one driving the other.

2.2.2. Data Collection

Once goals are clear and hypotheses have been determined, it should be fairly obvious as to the general types of study designs to be employed, although more detailed questions will persist throughout the entire data collection process. Two major questions remain to be answered: *(1)* which array will be used to generate data and *(2)* which data source(s) will be used.

2.2.2.1. SELECTION OF ARRAYS

Technical- and business-related issues normally drive the decision on what array platform to use for predictive assessment. Because the best predictive genes are unknown before the reference database is built, a genome-wide array is optimal for training. Commercially available "oligo" arrays such as Affymetrix's GeneChip® arrays (Santa Clara, CA) and GE Healthcare (United

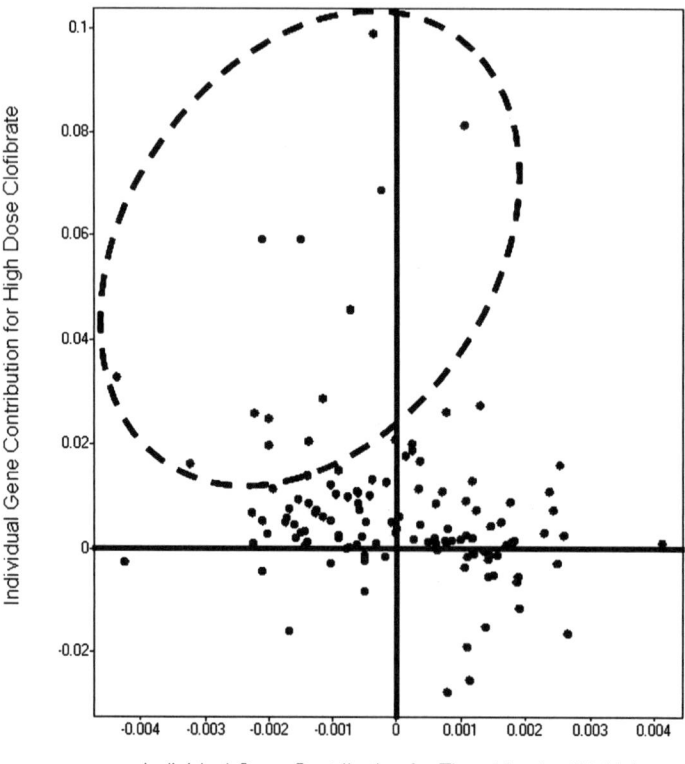

Fig. 2. Individual marker gene contributions from clofibrate within the liver perox-isome proliferation model. Each of the genes are displayed as a point on the graph and are each highly weighted with respect to their ability to discriminate the pathology. Positive gene contributions mean a "vote for" development of the pathology, meaning the gene is highly weighted and regulated in the direction of the pathology based on that gene's behavior within other peroxisome proliferation compounds. Negative gene contributions mean a "vote against" development of the pathology and regulation in the direction opposite of the pathology as observed in the reference database. The y axis shows a preponderance of positive contributions for a high dose of clofibrate, and the x axis shows a fairly random distribution of positive and negative contributions along with values closer to zero (note the smaller scale on the x axis relative to the y axis). The circled points are genes that are very highly contributing and should be the first genes further investigated for their relationship to mechanism.

Kingdom) CodeLink arrays as well as cDNA microarrays have all been successful in building robust training sets *(9,14,15)*.

Regardless of the array used, the relevant biology seems to be commonly captured in the form of regulated systems and pathways even though the same genes may not show significant coregulation across platforms. Therefore, a molecular toxicologist analyzing data from two or more platforms will generally come to the same conclusion cross-platform as far as the principal mechanistic trends. However, the individual gene-by-gene variability severely limits the ability to build a predictive model on a single platform and apply it to a completely different platform. To date, there are no public reports of cross-platform predictive modeling yielding an acceptable level of validation. However, there are unpublished reports of cross-platform predictive modeling from a proof-of-concept level of validation. Promising strategies of increased cross-platform compatibility using sequence-based commonality rather than clustering-based annotations in public databases have been reported, but these strategies do not completely alleviate the problem *(16,17)*. One strategy that has shown some promise is to build a subset of an already existing reference database on another platform so that the same exact RNA samples are run on both systems. If a sufficient number of treatment conditions are run on both platforms such that variability and response can be properly modeled across the two, then a predictive model trained on the primary database may be successful in predicting test samples on the other platform with minimal decrease in accuracy. Obviously, stringent validation of a cross-platform "conversion" is necessary before it is fully implemented.

2.2.2.2. SOURCES OF TOXICOGENOMICS DATA

Data sources for predictive modeling can be separated into five categories (with accompanying citations of examples):

1. Private, where data is collected within a single organization and not distributed externally.
2. Commercial, where data is collected within a single organization and marketed externally *(9)*.
3. Consortia, where data is collected by multiple institutions and shared among all participants *(18)*.
4. Government, where data is collected by one or more institutions and provided by a publicly funded institution *(19–21)*.
5. Public data repository, where data is deposited by multiple institutions with no specific common focus or goal *(22)*.

Each type of database is structured differently on an economic basis and has advantages and disadvantages from a scientific and data-quality perspective that are beyond the scope of this review. The proper choice will most likely depend on the resources available for in-life experimentation and microarray processing as well as cash available, but should also depend highly on data-quality concerns. For instance, the degree of data quality is much greater for private or commercial databases than for the public data repositories, as private and commercial enterprises incorporate process flows and controls for all aspects of decision-making and data collection. Public data repositories normally suffer from a lack of clear direction, as they accept data from multiple institutions with many causes, and from lack of robust quality control.

2.2.3. Data Parsing

Data parsing simply defines the way in which data is used to train the model. For example, if a two-class, discrete classifier is being constructed, each sample in the training set must be assigned to one of the two classes (i.e., binary assignment) or, if a discrete classifier is being constructed with more than two classes, each sample must be assigned in some way to one or more of the classes. Once the classes have been defined, the classification algorithm will determine one or more functions that optimize separation of the classes. Classifiers based on a continuous metric can also be constructed, where each sample is not assigned a "class 1 versus class 2" membership but is instead used along with all of the other samples to create a function or line that can be regressed to true test cases. The position within the function may be assigned by any number of parameters, including gene expression data itself.

Although the concept of data parsing is intuitive, the actual exercise of parsing samples based on knowledge of toxicology for multiple compounds is fraught with difficulties that need to be addressed. A complete understanding of the goals of the models and the hypotheses that are being tested by the model will aid in this process.

2.2.3.1. TRUE TOXICITY POTENTIAL AMBIGUITY

Compounds are known to cause different toxicities across experiments. Even when basic animal husbandry variables are well controlled, the appearance and prevalence of certain pathologies can be inconsistent for the same compound treatment. Literature research can help to quantify the amount of variability of toxicities for some, but not all, toxicants. Therefore, compounds that are well documented to cause the same types of toxicities in the same dose ranges in a consistent nature serve as the best training set.

2.2.3.2. MULTIPLE END POINTS

A large number of compounds cause more than one type of toxicity for a target organ. For any combination of two or more liver pathologies, for example, many compounds within the respective training sets for each individual pathology end point may be in both sets. If the training sets for models based on end point are highly correlative, their results will be highly correlated and the system will not be able to effectively discriminate between the intended end points. Two methods of attenuating this effect are presented.

First, models can be built such that end points are modeled based on training compounds that cause a single pathology only. This strategy basically removes the effect of correlated end points entirely, and all predictor genes can be thought of as true discriminators of the intended classification. However, the training sets are usually limited by this procedure, as very few well-known toxicants cause a single pathology or induce development of other toxicity end points. This limitation in training set size may hinder generalization of the model to the larger population of compounds. However, the strategy may be worth the time and resources due to the "pure" response of the genes. In validation of this strategy, care must be taken to ensure that compounds that cause multiple pathologies are successfully predicted by all relevant "pure" models, as synergistic effects mediated by gene expression for the multiple end points may decrease sensitivity.

Alternately, one can build models such that any compound that causes a toxicity end point, regardless of other end points within the organ system, are used in the training set. Care must be taken in this case to ensure that the intended end point is the prevalent one and to know what other end points are represented and at what level in comparison to the principal.

2.2.3.3. Incidence and Severity Versus All-or-Nothing Toxicity

Compounds are not "all toxic" or "all nontoxic" to any one organ system. Toxicity is an inherently continuous metric, defined by a series of descriptions, most of which are not easily quantifiable with respect to an animal model. Therefore, in assigning membership to classes in a discrete all-or-nothing fashion, qualifications need to be made concerning the severity and incidence of the toxicity in the animal and organ system for which predictions are being output. Not all compounds cause the same degree of toxicity in terms of acute or chronic effects and not all compounds adversely affect all individuals, most notably in humans where idiosyncratic toxicities are a large concern. Efficacious and adverse dose ranging is the prevailing factor that necessitates a rather laborious process of establishing dose, incidence, severity, and multiple end-point relationships. These relationships lend themselves better to a continuous class membership than to a binary one. However, experimental results are too inconsistent and information about compounds too scarce to properly establish these relationships for all but the most well-characterized few. Despite these issues, predictive models have done a relatively good job of providing continuous scores for test compounds using classifiers that are based on discrete, binary parsing. The key may lie in only using those compounds with well-established toxic effects.

2.2.3.4. LOW DOSES RELATIVE TO BENCHMARKED HIGH DOSES

The dose parameter as it relates to model training may be clear if a benchmark of toxicity is established, but it is not clear how doses lower than this benchmark should be used to train a model. A dose within the efficacious range of a drug may cause gene regulation associated with the beneficial aspects of the drug but also may harbor regulation associated with the toxic response at more subtle levels. This is particularly evident in organs where toxicity and pharmacological responses are both prevalent such as the heart. Because no universal assumption can be made concerning the dose-response curve as it relates to toxicity using gene expression data, low doses should probably be excluded from model-building unless a toxic response is observed within the gene expression data by in-depth mechanistic analysis. Once the model is built, various doses of test compounds can be used to show the types of dose-response curves resulting from the assay and can be compared to classic indicators of toxicity.

2.2.3.5. TREATMENT OF DATA FROM MULTIPLE EXPOSURE TIME POINTS

Within each toxicogenomics study, there are usually multiple time points with data ranging from very short exposure times such as 1 hour to 2 weeks or more of exposure. Depending on various aspects of the drug and the route of administration, genes related to the toxic response are regulated differentially over time, where some time points seem to have little to no gene expression changes and other time points have robust changes, measured by numbers of genes and the level of change. There is no single way to treat multiple time point data in building models. Gene Logic has attempted to build models by two methods. In the first method, only time points that harbor robust gene expression changes are used to train. The method of defining the limits of a "robust gene expression change" are outside the scope of this review but involves an assessment of variability of each time point relative to the rest of the reference database. Although this method yielded acceptable models, the second method, which uses all time points in the training set regardless of the number of gene expression changes, results in greater sensitivity. This increase in sensitivity is likely due to the model being trained on more subtle changes found within the time points that seemed to be of little value at first glance. When combined with a large reference database, genes of seemingly small changes can result in highly discriminative and consistent markers.

2.2.4. Statistical Methodology

The training of models necessitates a variety of decisions concerning the application of numerous statistical procedures. A wide variety of unique combinations of techniques have been researched and implemented for toxicogenomics and other biological systems such as tumor classification *(5,8–10,13,15,23–26)*. The toxic response seems to be a robust and reproducible phenomenon as captured by whole-genome microarrays, and a large number of methods have been used to build highly successful models. Therefore, the most important factors for the statistician to devote time and effort toward are not the exact implementation of predictor gene selection or classification algorithm but in ensuring that the model can be robustly generalized beyond the training set and that the model outputs are interpretable.

Although multiple methods work, the statistician or researcher must still decide on model parameters within three areas: *(1)* data preprocessing, *(2)* feature selection methods and criteria, and *(3)* classification algorithm.

2.2.4.1. DATA PREPROCESSING

Data preprocessing includes normalization of arrays to facilitate comparison of two or more experiments, identification and treatment of outlier arrays, and identification and treatment of arrays that are of poor quality. All of these steps have previously been reviewed, and each continues to generate a large degree of discussion on which methods are most reliable for a variety of study designs and platforms *(27–31)*, leaving the final decision of which normalization, outlier, and quality control methods and filters to use as a situational exercise for the statistician.

Once the generation of the data set is completed, one must decide the base data type(s) upon which the model will be built. This base data type will also be used for incoming test samples. The statistician must determine if absolute signal intensity or relative regulation will be used. The advantage of relative regulation such as fold changes of treatment groups relative to time-matched vehicle controls over signal intensity is that it theoretically reduces the effect of confounding factors. For instance, a comparison of rats that are sacrificed first thing in the morning to rats that are sacrificed in the evening will likely yield many more differences than if they were first normalized to time-matched vehicle baselines, as the circadian rhythm has been shown to be a prevalent effect captured by molecular regulation *(32–34)*. Although fold change–based normalization may not get rid of all confounding effects, it will help in this endeavor as long as time-matched vehicle controls are used that are generated in conjunction with the treatment samples. Vehicle controls that are either not time-matched or were not generated with the treated samples do not offer the same degree of advantage in using a relative data type as the basis for the model. Fold change–normalized data relative to a pooled group of universal vehicle standards, for instance, will be sensitive to more confounding effects than if time-matched vehicle controls from the same study were used.

2.2.4.2. Feature Selection Methods and Criteria

As mentioned previously, a reduced number of predictors primarily increases interpretability of modeling results, whereas an increased number of predictors normally increases accuracy and generalizes better to the universe of toxicity mechanisms induced by compounds. The hunt for models of the lowest gene size and highest accuracy is elusive at best, as the perfect combination of genes may be achieved only after a combinatorial exercise over tens of thousands of candidates. This combinatorial exercise is likely to yield gene combinations that are overfit and do not generalize well, especially if combined with a relatively low number of compounds on either side of the training set.

The genes responsible for discrimination between toxicity end points may be known to take part in a mechanism based on previous studies of transcript regulation, protein expression, and/or enzymatic activity. However, the majority of discriminative genes either are not annotated as to function or are known functionally but have not previously been reported to contribute to the mechanism. The identification of previously unknown predictive toxicity markers using a modeling-based approach allows one to further research and refine mechanistic knowledge, although confirmation of such events on a potentially less noisy platform such as Q-RT-PCR is suggested.

Gene Logic's predictive modeling program has yielded several unpublished conclusions concerning marker gene set characteristics. First, homogeneity of the gene expression response allows for reduced gene list size. For fibrate compounds that induce peroxisome proliferation in the liver, the homogeneity of the response is very high, allowing for predictive marker sets in the dozens. However, the homogeneity of the response for liver enlargement is much less, necessitating marker sets in the hundreds for successful modeling of test case inducer compounds. Second, most good markers are not currently known to contribute to the mechanism being assayed. Third, interpretive tools that allow one to view the contributions of the individual genes to the overall prediction are a necessary part of the assay results for both confirmation and discovery purposes.

Genes that adhere to the following requirements should be identified for use in a predictive model:

- Cohesive regulation across the toxicity or type of toxicity being predicted.
- Insensitivity to a variety of confounding factors, either stemming from biological background events germane to the model organism and inherent study design (e.g., gender, strain, housing, pharmacological effects, etc.) or measurement platform (e.g., noisy signals).
- Consistent regulation over time or, at the least, regulation at time points most commonly used in preclinical assays.
- Regulation that is either *predictive* or *correlative* of observable toxicity end points.

For *predictive* events, the regulation either precedes an observable toxicity within the same species or mirrors events that occur in another species (e.g., common mechanisms that may be launched in both human and rat, but only result in toxicity within the human due to some downstream attenuation of the rat response).

For *correlative* events, the regulation coincides with the observable toxicity indicator.

Although predictive events are preferred, there is some value in correlative genes being included within a model. For instance, it is difficult to explain to scientists not familiar with toxicogenomics why, for a particular compound, a model does not show a positive result for a time point at which toxicity is obvious.

Most gene selection methods incorporate a filtering step that chooses sets of genes that discriminate two or more populations. These include *t*-tests, analyses of variance (ANOVAs), correlation measures, linear discriminant analysis, and so forth. Although highly intuitive and simple to use, these methods are not able to quantify cross-feature dependencies and information redundancy *(35,36)*. Multivariate methods are usually more computationally intensive but identify predictors that optimally contribute information to the model that it did not receive from the previously selected features. These include forward and backward selection, classification trees, and so forth. Researchers have also used combinations of methods to provide more robust predictive sets specific for their data *(37)*. For most toxicogenomics applications, any number of these methods can result in highly accurate models. In ensuring that the estimated success of the model can be generalized to the universe of compounds, the overall number of good marker genes is more important than the method used.

2.2.4.3. CLASSIFICATION ALGORITHMS

There are a large number of classification algorithms that can be successfully used to build a toxicogenomics predictive model. Each algorithm has inherent limitations and makes assumptions about the data used to build and test the model. The algorithm of choice for any number of data sets should take into account characteristics such as sample size, data distribution, available processing resources, and potential to overfit. Despite the wide variety of options, almost every class of technique has been tested and has demonstrated success based on all types of gene expression data, including toxicogenomics data. A general review of these methods are beyond the scope of this article and have been extensively discussed elsewhere (http://www.statsoftinc.com/textbook/stathome.html and *(9,35)*.

As a part of the entire predictive modeling process, the classification algorithm of choice is one of the lesser factors of importance given a robust gene expression response. For robust responses, improvements in success rates based on classification algorithm are likely to yield only a few percentage points difference and are likely not to be successfully generalized in the long run. Ensuring that the feature selection and classification algorithms are synergistically combined to yield a "generalizable" classifier is much more important than increasing the success rate by a few percentage points.

2.2.5. Model Validation

Once a model has been built and its success rates determined as a function of the training set, it is important to use multiple test sets to better estimate success over the entire population. Toxicogenomics data is more difficult than other types of data to successfully validate primarily due to three factors: *(1)* database structure, *(2)* size of the universe of compounds and mechanisms, and *(3)* ambiguity concerning true toxicity potential of compounds. In addition to these three factors, another issue that requires attention is the treatment of feature reselection during cross-validation.

2.2.5.1. DATABASE STRUCTURE

As shown in **Fig. 3A**, a typical toxicogenomics experiment has a large degree of structure associated with it. Typically for each compound, multiple time points and doses may be represented and used within the training set. Therefore, a high degree of dependency exists between the different treatment groups, as a significant number of toxicity response genes are known to exhibit consistent regulation over time and across dose. In addition, multiple replicates are usually found within each treatment group, yielding a large degree of dependency between samples. One method of removing replicate variability is to use summarized expression values across the replicates for each treatment group. This step does not remove the estimated high degree of dependency between time and dose parameters, however.

Within the framework of the database structure, a random sampling validation is insufficient to treat each removed observation as independent from the training observations. Calculated success rates are vulnerable to overestimation within the removal of random samples, both at the level of individual replicates (**Fig. 3B**) and summary values. Multiple iterations of random validation will not solve this problem and will only give a more stable overestimated value. The optimal method of validation where the training set is independent of the test set is to remove all samples from an entire

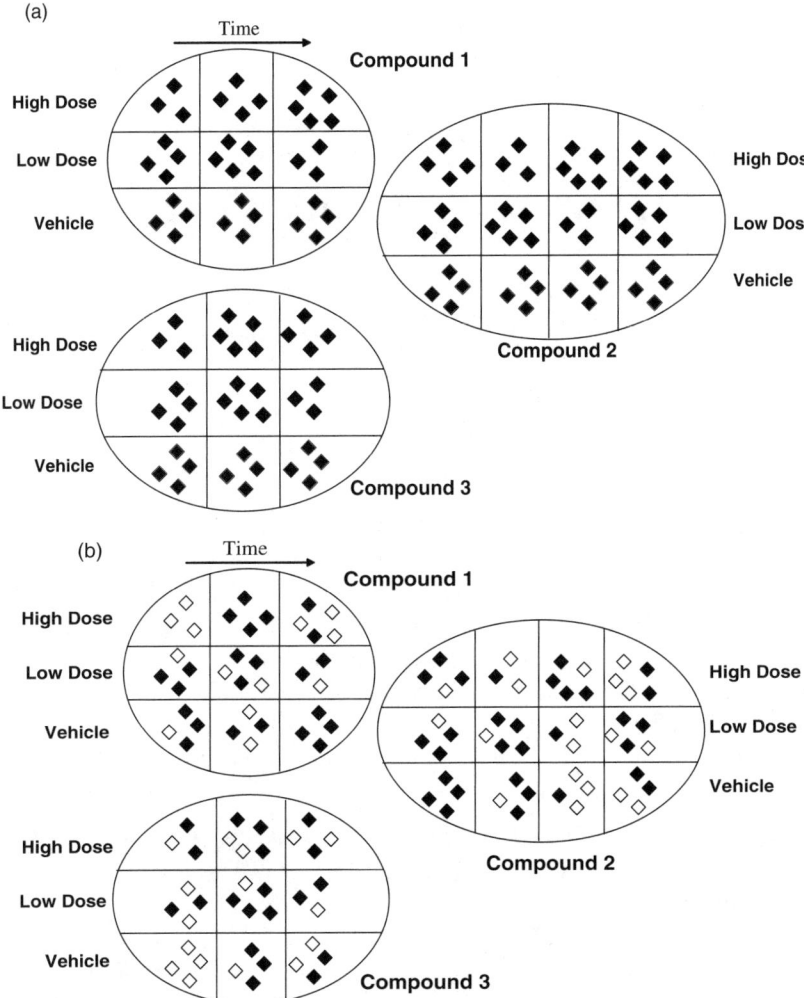

Fig. 3. Cross-validation schemes within a model based on toxicogenomics database structure. (A) Structure of a toxicogenomics database showing multiple time points, doses, and treatment group replicates, where each solid diamond represents a sample used to build the predictive model. (B) Random "sample split" validation scheme where a certain percentage of samples are randomly removed from the training set, represented by open diamonds. Dependencies exist between treatment group replicates, across time, and between doses. Success rate estimations based on this scheme will be inflated, most likely to a large extent. (C) Compound drop validation scheme, where an entire compound's data has been removed from the training set. This validation scheme ensures independence of the test set and training set during cross-validation and will provide a good estimation of the true success rate.

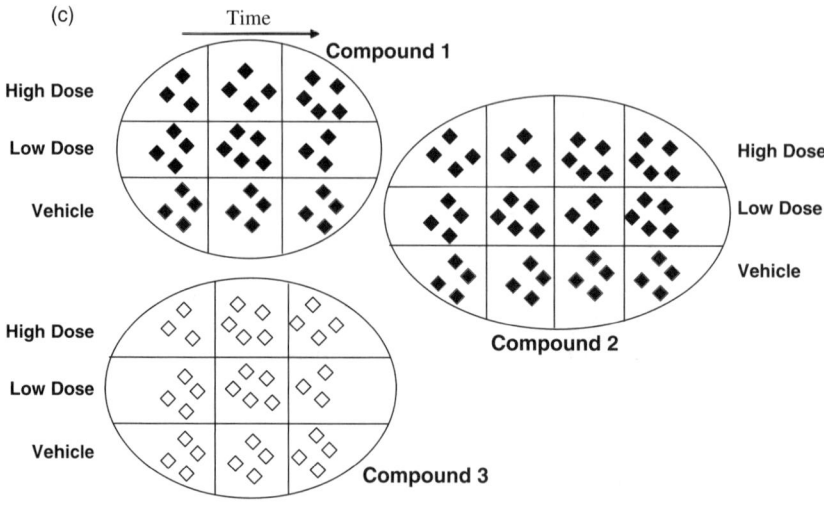

Fig. 3. (Continued)

compound's experiment(s), retrain the model, and apply the model to the removed experiment (**Fig. 3C**). This robust method is closest to the way in which the model will be ultimately implemented and achieves optimal independence of the test set.

2.2.5.2. FEATURE RESELECTION

In proper validation of models, the predictor variables need to be reselected after the test portion of the data has been removed. In the *compound drop* paradigm described above, this requires that genes be reselected using the same criteria as was implemented in defining the full model after each compound's data is dropped. This removes any bias in success rates based on the removed compound still contributing to the selection of genes that optimize its prediction. Estimates of the selection bias on predictive modeling using microarrays has previously been reported *(38)*, and this bias is known to exist for toxicogenomics modeling based on several unpublished analyses.

2.2.5.3. UNIVERSE OF COMPOUNDS AND TOXICITY MECHANISMS

The results of the model validation exercise will yield an estimate of success based on the compounds and mechanisms represented in the training set. However, it is unclear if these training compounds and their associated toxicity mechanisms are truly representative of all mechanism space for a pathology or other toxicity end point. It is also unclear how many "representative" mechanisms exist for each target organ's toxicity, how many representatives exist within the training set, which mechanisms can or cannot be captured by gene expression measures, and how many of each representative mechanism would be necessary to build a model robust for the entire population. As an example, liver peroxisome proliferation seems to be very homogeneous in its observed gene expression response, and the number of compounds needed for the training set and for successful prediction of a test set is rather small (i.e., 5 to 10). Alternately, liver necrosis seems to be rather heterogeneous, and the number of compounds needed for the training set is most likely beyond 20. This phenomenon could be due to the heterogeneity of the gene expression responses that result in necrosis and the lack of representatives for each of the mechanisms resulting in necrosis.

This point is highly theoretical but important in establishing that in estimation of success rates for a model, one can only observe what is available. If that sampling is an extremely small representation of the total population, then any model is suspect, no matter how high the calculated success rate. As it is currently unclear what the total number of mechanisms is for any particular type of toxicity, the prediction of the toxic response using gene expression should be qualified as one of these cases.

2.2.5.4. TRUE TOXICITY POTENTIAL AMBIGUITY

As the true number of representative mechanisms for any toxicity end point is currently unknown, there is a need to continually test the model using agents of known toxicity. However, there is often a great deal of ambiguity associated with the true toxicity potential of a compound, as compounds are capable of inducing various individual-specific responses. This is true within the animal being measured such as the rat but is even a larger issue in humans, where genetic variation is greater than that of animal model systems. Validation of toxicogenomics predictive models, then, is not as quantifiable as one would like on an ongoing basis. Most validation occurs in the form of anecdotal interpretation, where objective thoughts on confirmation of known toxicities and conjecture concerning potential toxicities relative to modeling output prevail.

2.2.5.5. VALIDATION SUMMARY

The true success of gene expression–based models is likely to only be realized after a decade or more of testing within healthy drug development pipelines across several pharmaceutical companies. The validation estimates of current models, while promising, should be considered as proof-of-concept, as the content currently available for estimating success rates is insufficient to estimate universal performance.

In cases where a model is built for a toxicity end point of interest but fails to yield high success rates using a cross-validation or true test case approach, the following questions should be asked:

1. How similar is the data across the training set compounds within each class at the global gene expression level?
2. How similar is the data across the training set compounds within each class using genes known to be involved with their associated mechanisms?
3. Are previously unknown mechanisms for the predicted end point still being discovered using novel compounds or compound series?
4. Are the known mechanisms for the predicted end point complex or simple?
5. Do the known mechanisms for the predicted end point rely on changes at the transcriptional level or at other levels such as posttranslational modifications? A good case example of an adverse event that does not seem to manifest itself robustly in gene expression regulation is QT prolongation.

If the answers to these questions are unknown, then more research should be done into the mechanisms involved for each of the compounds composing the class of interest. Mechanistic signatures of a few dozen or less genes known to be associated with the mechanism, though not robust for modeling purposes, are excellent for providing additional information about a compound's profile. For a model that does not seem to work as specified, mechanistic signatures related to the development of the end point will provide more usable information.

Finally, it is ill-advised to "force" a model that adheres to exact specifications by finding a solution space that is exceedingly difficult to engineer. If a model yields low success rates after trying a couple of types of gene selection and classification algorithms and there is no immediate suggestion of improvements that can be made regarding data parsing or collection strategies, one should seek a more appropriate avenue for research of the hypothesis, such as a mechanistic signature analysis using genes known to contribute to the toxicity response.

2.2.6. Model Implementation

The final step in the modeling process reviews researcher access to modeling results and the types of results that are presented. As the model is based on arithmetic combinations of weighted gene statistics, the model will always suffer from a "black box" reputation. Ways of alleviating the black box include intuitive interface–based access to models, full disclosure of predictor gene behavior, full disclosure of model training, methods used, and validation results, and intuitive visuals that relay confidence in results.

2.2.6.1. INTUITIVE MODEL ACCESS

One of the key components for acceptance of the model by biologists is to allow the researchers themselves to access the model, navigate the output results, and gain some understanding of its underlying behavior and metrics. This can be achieved without an intuitive interface but severely decreases the size and skill sets of the audience, as it necessitates programming capabilities, usually within a specialized statistical programming language. The implementation of an interface requires work from software developers and statistical programmers. Very few products are currently available that allow this level of access, but as proof-of-concept experimentation further validates the technology, a demand for maximizing access to researchers is increasingly likely.

2.2.6.2. PREDICTIVE MARKER GENE BEHAVIOR

The requirements for a predictive model system should include a means of accessing the contributions for each individual predictor gene to a modeling result. Further discussion of why this is necessary and an example of one way in which this can be implemented are provided in **Section 2.2.1.5**.

2.2.6.3. MODEL TRAINING, METHODS, AND VALIDATION

Both toxicologists and statisticians who use the resulting model system will require knowledge of which toxicants and negative controls were used to train each model and how. Documentation of how models were trained should at least include the following for each model:

- Compound identifier/name
- Compound class membership for each model
- Compound training scores according to dose and time points of sacrifice, route of administration, and so forth
- A definition of each model in terms of goals/hypotheses, caveats, and known issues
- Gene selection method used
- Classification algorithm used
- Validation scheme and rules for determination of true positive/negative rates
- Threshold used for binary determination of class
- Method used for calculation of probabilities for continuous determination of class
- Prediction of each compound under the validation scheme for each time point and dose

2.2.6.4. INTUITIVE VISUALS

The goal of predictive modeling is to separate two or more populations of samples with some level of probability estimated for each model output score. These probabilities can be calculated in a variety of ways and should be based on the success of the model in separating classes based on true test cases and/or rigorous validation schemes. In addition to the reporting of a numeric probability, a much more intuitive confidence is based on the behavior of each test sample from the entire experiment as visualized by side-by-side comparison with compounds of well-known toxicity.

An example of this type of plot is shown in **Fig. 4** (*see* Color Plate 8) as implemented within Gene Logic's predictive modeling system. The y axis

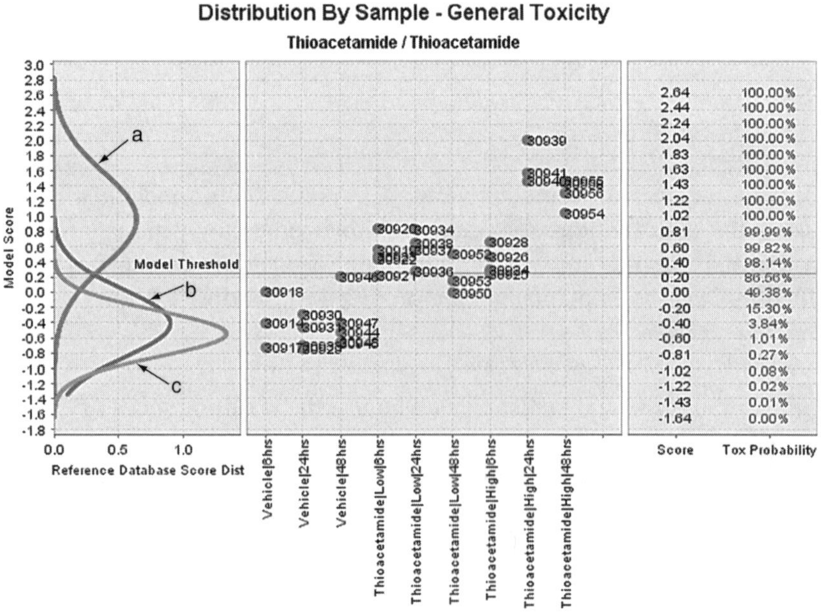

Distribution By Sample - General Toxicity

Thioacetamide / Thioacetamide

Fig. 4. Model scores for two doses of thioacetamide or vehicle-treated samples at 6-, 24-, and 48-h exposures. The plot from Gene Logic's ToxExpress System database is provided as an example of a diagnostic visual of toxic response based on a model system. A full description of the plot is provided in the text in **Section 2.2.6.4**. In short, the leftmost portion of the plot shows the behavior of reference database known toxicant-treated samples (indicated by arrow 'a'), known non–toxicant-treated samples (indicated by arrow 'b'), and vehicle-treated samples (indicated by arrow 'c') as a function of the output model score. The model in this case is assessing general liver toxicity of any type. The rightmost portion of the plot shows the probability of toxicity as calculated based on the overlapping distributions seen in the leftmost portion. The middle portion of the plot shows the experimental samples plotted relative to the model score. Each treatment group is divided by the x axis and individual samples within each treatment group are labeled. The horizontal line is the binary threshold above which a response should be deemed "toxic." The continuous nature of the modeling output is evident and provides critical information such as intragroup versus intergroup variability, dose and time trends, proximity of response to the binary threshold, and relative confidence based on reference database behavior. All of this information is critical to providing correct interpretation for a modeling result. (*see* Color Plate 8).

is constant throughout the plot as the output continuous score of the model. A horizontal line has been drawn to indicate the predefined threshold above which a sample should be considered positive for the model and below which a sample should be considered negative to facilitate a binary classification.

The x axis is different for each of three sections of the plot:

- The leftmost portion of the plot shows the behavior of compounds of well-known toxicity potential as shown by a frequency distribution, where frequency is plotted on x relative to the continuous model output score plotted on y. The frequency distribution is based on individual samples' output model scores where the distributions are coded as to whether the component samples are part of the two principal classes (i.e., toxic vs. nontoxic, necrosis vs. not necrosis, etc.). One can quickly determine the degree to which the model successfully separates samples of the two populations based on well-studied compounds and the consequence of being above or below the horizontal threshold line. The normal distribution is somewhat of a simplification for visual purposes but is close to what is observed within this example system.
- The rightmost portion of the plot shows output model score intervals and their associated probabilities. In this example, a probability of 100% means that no reference negative control or vehicle sample ever achieved that high of a score. Likewise, a probability of 0% means that no reference toxicant ever achieved that low of a score. Intermediate probabilities are based on the overlap of distributions between the toxic vs. nontoxic distributions as calculated by standard density function methods.
- The middle portion of the plot shows the behavior of the test samples relative to treatment group and continuous model score. Each sample is plotted relative to its output model score within its treatment group **Fig. 4**. Alternately, summary visuals such as box-and-whisker plots can be implemented to show group behavior in terms of mean, median, interquartile range and outliers are shown. From these plots, one can quickly determine confidence in prediction outputs based on several factors including:

1. Overall trends in model scores relative to the treatment groups. This allows for quick observation of dose and time responses/trends.
2. Intragroup variability versus intergroup variability. One can determine baseline variability from vehicle or control treatments and can compare to treatment-based variability.
3. Resulting model score relative to distributions of scores of well-known toxicity potential contained in the leftmost portion.
4. Resulting model score relative to calculated probabilities in the rightmost portion.

Accurate interpretation of modeling results is critically dependent on the information provided within this plot. A binary assessment of class 1 versus class 2 will only yield the most basic information and will not give any estimate of confidence. In fact, within Gene Logic's predictive modeling system, true positive/negative rates with respect to a binary threshold are only provided because researchers expect this sort of information. However, in assessing the toxicity potential of one or more actively developing drugs, the continuous metric with associated probabilities and experimental variability is implemented and relied upon far more than the binary assessment.

Finally, all predictive models need to provide probabilities as a key component of the output. It is insufficient to present an output model score and characterize it as "strong," "high," "weak," or "low" in the absence of a calculated probability of toxicity, as such characterizations provide very limited indication of significance.

3. Conclusion

Predictive toxicogenomics using microarray data over a large number of compounds and treatment conditions offers a great deal of promise in more completely characterizing the toxicity potentials of developing drugs and molecules. However, all of the above listed issues must be acknowledged even if the researcher does not know the best way to handle or implement any particular one. In viewing model-building and implementation as a much larger process, one sees the complexity inherent in the entire system.

Of particular importance is the realization that most issues addressed in this chapter are the same issues that classic toxicology has dealt with for decades. Items such as dose response, cross-species responses, differential time responses, multiple toxicity end points caused by single compounds, variability of results across individuals, incomplete estimates of incidence and severity across dose, and generalization of results to all compounds and mechanisms are nothing new to toxicology.

In conclusion, predictive modeling based on gene expression data is a rather new tool that can be used to provide additional confidence in moving forward with a developing compound or can be used in conjunction with current assays to justify a safety concern. Given a large number of potential compounds in a drug development pipeline that need to be narrowed to only a few viable candidates, it is another screening tool based on its ability to quantify molecular changes associated with the toxic response. It is also valuable in identification of potential toxicity biomarkers. As long as researchers understand the limitations discussed and are enabled to appropriately question the entire process of model-building as defined above, predictive modeling in toxicogenomics can provide more powerful information in making drug development decisions than is currently realized.

References

1. Hamadeh, H.K., Bushel, P.R., Jayadev, S., Martin, K., DiSorbo, O., Sieber, S., et al. (2002) Gene expression analysis reveals chemical-specific profiles. *Toxicol. Sci.* **67**, 219–231.
2. Waring, J.F., Jolly, R.A., Ciurlionis, R., Lum, P.Y., Praestgaard, J.T., Morfitt, D.C., et al. (2001) Clustering of hepatotoxins based on mechanism of toxicity using gene expression profiles. *Toxicol. Appl. Pharmacol.* **175**, 28–42.
3. Huang, Q., Jin, X., Gaillard, E.T., Knight, B.L., Pack, F.D., Stoltz, J.H., et al. (2004) Gene expression profiling reveals multiple toxicity endpoints induced by hepatotoxicants. *Mutat. Res.* **549**, 147–167.
4. Kramer, J.A., Curtiss, S.W., Kolaja, K.L., Alden, C.L., Blomme, E.A., Curtiss, W.C., et al. (2004) Acute molecular markers of rodent hepatic carcinogenesis identified by transcription profiling. *Chem. Res. Toxicol.* **17**, 463–470.
5. Bushel, P.R., Hamadeh, H.K., Bennett, L., Green, J., Ableson, A., Misener, S., et al. (2002) Computational selection of distinct class- and subclass-specific gene expression signatures. *J. Biomed. Inform.* **35**, 160–170.
6. Roberts, R.A., Michel, C., Coyle, B., Freathy, C., Cain, K., and Boitier, E. (2004) Regulation of apoptosis by peroxisome proliferators. *Toxicol. Lett.* **149**, 37–41.
7. Ramos, K.S., Chacon, E., and Acosta, D., Jr. (1996) Toxic responses of the heart and vascular systems. In: *Casarett and Doull's Toxicology: The Basic Science of Poisons* (Klaassen, C.D., ed.), McGraw-Hill, New York, pp. 487–527.
8. Steiner, G., Suter, L., Boess, F., Gasser, R., de Vera, M.C., Albertini, S., and Ruepp, S. (2004) Discriminating different classes of toxicants by transcript profiling. *Environ. Health Perspect.* **112**, 1236–1248.
9. Porter, M.W., Castle, A.L., Orr, M.S., and Mendrick, D.L. (2003) Predictive toxicogenomics. In: *An Introduction to Toxicogenomics* (Burczynski, M.E., ed.), CRC Press, Boca Raton, pp. 225–259.
10. Hamadeh, H.K., Bushel, P.R., Jayadev, S., DiSorbo, O., Bennett, L., Li, L., et al. (2002) Prediction of compound signature using high density gene expression profiling. *Toxicol. Sci.* **67**, 232–240.
11. Pahl, A., and Brune, K. (2002) Stabilization of gene expression profiles in blood after phlebotomy. *Clin. Chem.* **48**, 2251–2253.
12. Ellinger-Ziegelbauer, H., Stuart, B., Wahle, B., Bomann, W., and Ahr, H.J. (2004) Characteristic expression profiles induced by genotoxic carcinogens in rat liver. *Toxicol. Sci.* **77**, 19–34.
13. Ellinger-Ziegelbauer, H., Stuart, B., Wahle, B., Bomann, W., and Ahr, H.J. (2005) Comparison of the expression profiles induced by genotoxic and nongenotoxic carcinogens in rat liver. *Mutat. Res.* **575**, 61–84.
14. Kier, L.D., Neft, R., Tang, L., Suizu, R., Cook, T., Onsurez, K., et al. (2004) Applications of microarrays with toxicologically relevant genes (tox genes) for the evaluation of chemical toxicants in Sprague Dawley rats in vivo and human hepatocytes in vitro. *Mutat. Res.* **549**, 101–113.

15. Natsoulis, G., El Ghaoui, L., Lanckriet, G.R., Tolley, A.M., Leroy, F., Dunlea, S., et al. (2005) Classification of a large microarray data set: algorithm comparison and analysis of drug signatures. *Genome Res.* **15**, 724–736.

16. Carter, S.L., Eklund, A.C., Mecham, B.H., Kohane, I.S., and Szallasi, Z. (2005) Redefinition of Affymetrix probe sets by sequence overlap with cDNA microarray probes reduces cross-platform inconsistencies in cancer-associated gene expression measurements. *BMC Bioinformatics* **6**, 107.

17. Mecham, B.H., Klus, G.T., Strovel, J., Augustus, M., Byrne, D., Bozso, P., et al. (2004) Sequence-matched probes produce increased cross-platform consistency and more reproducible biological results in microarray-based gene expression measurements. *Nucleic Acids Res.***32**, e74.

18. Pennie, W., Pettit, S.D., and Lord, P.G. (2004) Toxicogenomics in risk assessment: an overview of an HESI collaborative research program.*Environ. Health Perspect.* **112**, 417–419.

19. Mattingly, C.J., Colby, G.T., Forrest, J.N., and Boyer, J.L. (2003) The Comparative Toxicogenomics Database (CTD). *Environ. Health Perspect.* **111**, 793–795.

20. Tong, W., Cao, X., Harris, S., Sun, H., Fang, H., Fuscoe, J., et al. (2003) ArrayTrack—supporting toxicogenomic research at the U.S. Food and Drug Administration National Center for Toxicological Research. *Environ. Health Perspect.* **111**, 1819–1826.

21. Waters, M., Boorman, G., Bushel, P., Cunningham, M., Irwin, R., Merrick, A., et al. (2003) Systems toxicology and the Chemical Effects in Biological Systems (CEBS) knowledge base. *EHP Toxicogenomics* **111**, 15–28.

22. Edgar, R., Domrachev, M., and Lash, A.E. (2002) Gene Expression Omnibus: NCBI gene expression and hybridization array data repository. *Nucleic Acids Res.* **30**, 207–210.

23. Tsai, C.A., Lee, T.C., Ho, I.C., Yang, U.C., Chen, C.H., and Chen, J.J. (2005) Multi-class clustering and prediction in the analysis of microarray data. *Math. Biosci.* **193**, 79–100.

24. Golub, T.R., Slonim, D.K., Tamayo, P., Huard, C., Gaasenbeek, M., Mesirov, J.P., et al. (1999) Molecular classification of cancer: class discovery and class prediction by gene expression monitoring. *Science* **286**, 531–537.

25. Nguyen, D.V., and Rocke, D.M. (2002) Tumor classification by partial least squares using microarray gene expression data. *Bioinformatics* **18**, 39–50.

26. Li, T., Zhang, C., and Ogihara, M. (2004) A comparative study of feature selection and multiclass classification methods for tissue classification based on gene expression. *Bioinformatics* **20**, 2429–2437.

27. Nadon, R., and Shoemaker, J. (2002) Statistical issues with microarrays: processing and analysis. *Trends Genet.* **18**, 265–271.

28. Zien, A., Aigner, T., Zimmer, R., and Lengauer, T. (2001) Centralization: a new method for the normalization of gene expression data. *Bioinformatics* **17** (Suppl 1), S323–331.

29. Choe, S.E., Boutros, M., Michelson, A.M., Church, G.M., and Halfon, M.S. (2005) Preferred analysis methods for Affymetrix GeneChips revealed by a wholly defined control dataset. *Genome Biol.* **6**, R16.
30. Irizarry, R.A., Bolstad, B.M., Collin, F., Cope, L.M., Hobbs, B., and Speed, T.P. (2003) Summaries of Affymetrix GeneChip probe level data. *Nucleic Acids Res.* **31**, e15.
31. Kreil, D.P., and Russell, R.R. (2005) There is no silver bullet—a guide to low-level data transforms and normalisation methods for microarray data. *Brief Bioinform.* **6**, 86–97.
32. Desai, V.G., Moland, C.L., Branham, W.S., Delongchamp, R.R., Fang, H., Duffy, P.H., et al. (2004) Changes in expression level of genes as a function of time of day in the liver of rats. *Mutat. Res.* **549**, 115–129.
33. Kita, Y., Shiozawa, M., Jin, W., Majewski, R.R., Besharse, J.C., Greene, A.S., and Jacob, H.J. (2002) Implications of circadian gene expression in kidney, liver and the effects of fasting on pharmacogenomic studies. *Pharmacogenetics* **12**, 55–65.
34. Boorman, G.A., Blackshear, P.E., Parker, J.S., Lobenhofer, E.K., Malarkey, D.E., Vallant, M.K., et al. (2005) Hepatic gene expression changes throughout the day in the Fischer rat: implications for toxicogenomic experiments. *Toxicol. Sci.* **86**, 185–193.
35. Gerwien, R., and Hyde, C. (2003) Reducing the risk of drug discovery: the application of predictive modeling to preclinical development. *Preclinica* **1**, 247–252.
36. Inza, I., Larranaga, P., Blanco, R., and Cerrolaza, A.J. (2004) Filter versus wrapper gene selection approaches in DNA microarray domains. *Artif. Intell. Med.* **31**, 91–103.
37. Furlanello, C., Serafini, M., Merler, S., and Jurman, G. (2003) Entropy-based gene ranking without selection bias for the predictive classification of microarray data. *BMC Bioinformatics* **4**, 54.
38. Ambroise, C., and McLachlan, G.J. (2002) Selection bias in gene extraction on the basis of microarray gene-expression data. *Proc. Natl. Acad. Sci. U. S. A.* **99**, 6562–6566.

7

Bioinformatics: Databasing and Gene Annotation

Lyle D. Burgoon and Timothy R. Zacharewski

Summary

"Omics" experiments amass large amounts of data requiring integration of several data sources for data interpretation. For instance, microarray, metabolomic, and proteomic experiments may at most yield a list of active genes, metabolites, or proteins, respectively. More generally, the experiments yield active features that represent subsequences of the gene, a chemical shift within a complex mixture, or peptides, respectively. Thus, in the best-case scenario, the investigator is left to identify the functional significance, but more likely the investigator must first identify the larger context of the feature (e.g., which gene, metabolite, or protein is being represented by the feature). To completely annotate function, several different databases are required, including sequence, genome, gene function, protein, and protein interaction databases. Because of the limited coverage of some microarrays or experiments, biological data repositories may be consulted, in the case of microarrays, to complement results. Many of the data sources and databases available for gene function characterization, including tools from the National Center for Biotechnology Information, Gene Ontology, and UniProt, are discussed.

Key Words: bioinformatics; databases; functional genomics; gene annotation; protein interaction; toxicogenomics.

1. Introduction

Genomic experiments amass large data sets, requiring the integration of supportive information from several other sources, including the most recent gene annotations, to facilitate biological interpretation. Typically, after microarray analysis and identification of the most active, or significant, genes, further investigation must be performed to elucidate the relevant pathways and networks involved in eliciting the phenotype (e.g., toxicity). Thus, investigators must

From: *Methods in Molecular Biology, vol. 460: Essential Concepts in Toxicogenomics*
Edited by: D. L. Mendrick and W. B. Mattes © Humana Press, Totowa, NJ

integrate complementary information including gene names, abbreviations, and aliases for literature searches; cellular and extracellular locations; functional annotation; disease processes the gene participates in; and biological interaction data (e.g., protein-protein interactions) in order to comprehensively interpret the data. This information is oftentimes available in a variety of biological databases each serving a particular purpose or devoted to a specific data domain.

This chapter will describe six broad categories of databases as they relate to genomic data integration, including genome level, sequence level, protein level, functional annotation, protein interaction, and microarray databases (**Fig. 1**). Excluded are the metabolomic-related domains as reporting standards have yet to emerge, although they are in development (http://www.metabolomicssociety.org, http://www.smrsgroup.org). All of the databases exist in a complex data exchange continuum, where some databases rely entirely upon others for their information, others are nearly independent of the rest, and the remaining integrate data from several different levels.

In general, genome sequences, from databases such as Ensembl *(1,2)*, Entrez Genomes *(3)*, and the University of California Santa Cruz (UCSC) Genome Browser *(4)*, are the root of the universe. From these genomic templates, expressed sequence tags (ESTs) and cDNAs in GenBank *(3)* can be clustered together and associated with genes (i.e., UniGene; Ref. *3*), and exemplary, representative full-length sequences can be identified from GenBank and mapped back to locations in the genome (i.e., RefSeq; Ref. *3*). These genes are then annotated in databases such as Entrez Gene *(5)*, where functional information (Gene Ontology; Ref. *6*), and disease information (Online Mendelian Inheritance in Man [OMIM]; Ref. *3*) are integrated to provide a more comprehensive summary of the function of a gene. Similarly, elements from sequence-level databases (e.g., ESTs) can be associated with features printed on a microarray and related to a gene through its GenBank Accession number facilitating the annotation of gene expression profiles from the microarray experiments. Integration of genomic and proteomic data is also possible through sequence relationships, from the mRNA to the translated protein sequence. This facilitates further functional predictions, by providing protein domain and family information that may reveal functional characteristics, and protein-protein interaction data from databases such as BIND (Biomolecular Interaction Network Database) *(7)* and DIP (the Database of Interacting Proteins) *(8)*.

Currently, there is significant effort in the development of public repositories such as the Chemical Effects in Biological Systems Knowledgebase (CEBS) *(9)*, ArrayExpress *(10,11)*, and the Gene Expression Omnibus (GEO) *(12)* to facilitate data integration across multiple domains and to ensure public accessibility, as well as to support the development of comprehensive networks and computational models capable of predicting toxicity.

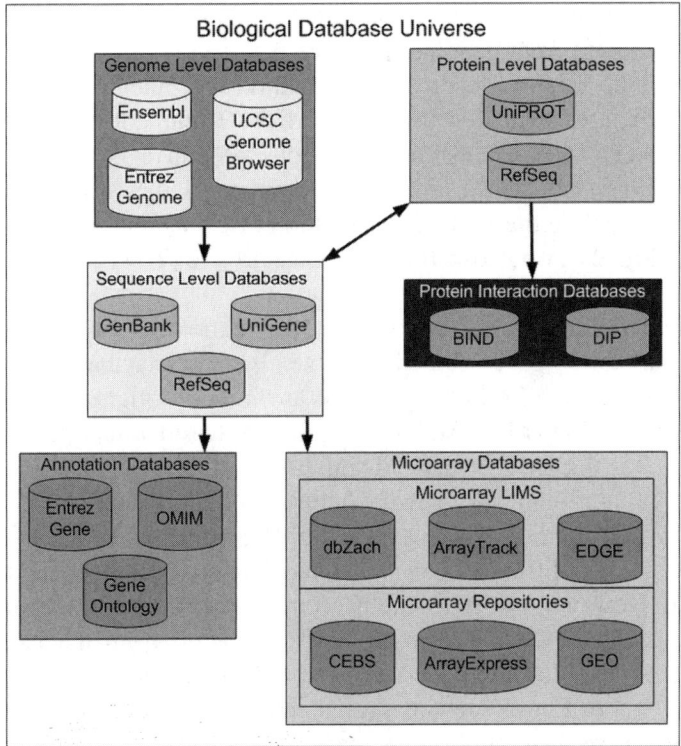

Fig. 1. The Biological Database Universe. Six biological database levels are depicted as they pertain to genomic data analysis and interpretation. Genome-level databases catalog data with respect to the sequence of the full genome. Sequence-level databases catalog sequence reads from cells, including genomic sequence and expressed sequence tags (ESTs). Annotation databases provide functional information about genes and their products. Protein-level databases provide information on protein sequences, families, and domain structures. Protein interaction databases provide interaction data concerning proteins, genes, chemicals, and small molecules. Microarray databases include local laboratory information management systems (LIMS) and data repositories. The arrows depict possible interactions between different database domains, where information from one level may exist in another to allow for cross-domain integration.

2. Genome-Level Databases

Genome-level databases manage, at the very least, genome sequence data. However, they differ in their integration of other types of data and often in their assignment of computationally defined genes. The three primary genome-level databases are the Ensembl database *(1,2)*, the Entrez Genomes database *(3)*, and the UCSC Genome Browser *(4)*. Each uses a different technique for predicting genes and gene structures (e.g., untranslated regions [UTR], regulatory regions, introns, and exons) from genome sequence data.

The Ensembl database uses several methods for the prediction of genes and gene structures that are biased toward the alignment of species-specific proteins and cDNAs, and using orthologous protein and cDNA alignments when necessary *(13)*. The use of the protein and cDNA alignments to the genome sequence facilitates the identification of exonic and intronic sequences and UTRs (**Fig. 2**). A putative transcription start site (TSS) can be obtained by defining the end of the upstream region.

The National Center for Biotechnology Information (NCBI) Entrez Genomes database annotates genes based on the RefSeq database of reference, exemplary sequences. RefSeq sequences are initially aligned to the genomic sequence using the MegaBLAST algorithm to identify genes; mRNAs and ESTs are aligned through MegaBLAST to identify additional genes (http://www.ncbi.nlm.nih.gov/genome/guide/build.html#contig).

The UCSC Genome Browser uses the NCBI genome builds for its annotation, thus, there are no differences between the human genome builds at UCSC and NCBI. However, prior to the December 2001 human genome freeze, the UCSC created its own genome builds, separate from the NCBI. Previously, the primary difference between the two methods was in their genome assemblies, where Entrez Genomes used sequence entries from the GenBank database to drive assemblies, whereas the UCSC Genome Browser used BAC clones and mRNA sequences, resulting in differences in the genome

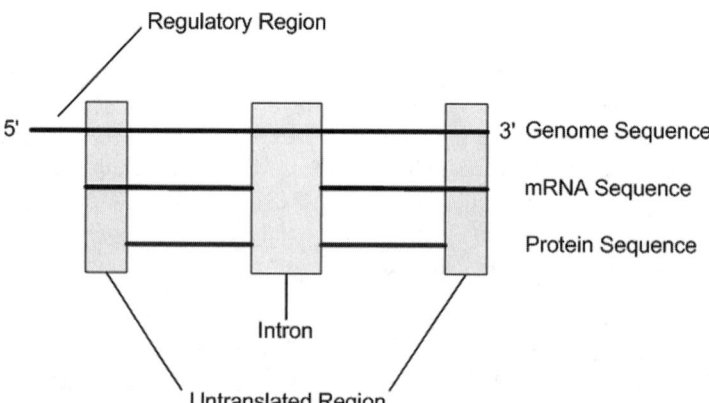

Fig. 2. Ensembl genome annotation. This simplified view illustrates the method used by the Ensembl genome annotation system for identifying gene structures, such as the untranslated region (UTR), exons, and introns, by combining genome, mRNA, and protein alignments.

assemblies *(14)*. For other genomes, such as the mouse (i.e., C57BL/6), rat (i.e., Norwegian Brown Rat), chimpanzee, rhesus monkey, and dog (i.e., Boxer), the UCSC uses builds from the respective genome authorities (see http://genome.ucsc.edu/FAQ/FAQreleases for further details).

To annotate the genome builds, NCBI uses the MegaBLAST algorithm for alignments to genomes, whereas the UCSC efforts use the BLAT (BLAST-like alignment tool) for alignment of mRNA, EST, and RefSeq sequences to the genome. This means that although both sources use the same build for the human genome (i.e., the NCBI genome build), there could still be differences in annotation (i.e., assignment of genes and functions to the genomic sequence). Assuming both use the same GenBank and RefSeq versions, differences may be attributed to the different alignment algorithms. In addition, the UCSC Genome Browser also incorporates gene predictions from other sources, such as Ensembl and Acembly *(4)*, and users can also upload their own annotations for display in the browser.

3. Sequence-Level Databases

Sequence-level databases manage EST and cDNA sequence read data. Some databases, such as GenBank and RefSeq, deal with these sequences directly, whereas others manage them on a larger scale, where multiple sequences are grouped together, as in UniGene. Generally, these databases provide the first level of annotation for microarray studies, as the sequences are directly represented on the microarrays as printed features.

When a sequence read is generated, it is generally submitted to the GenBank database and assigned a GenBank Accession Number, a unique identifier representing that sequence and is typically the most commonly used identifier for probes represented on cDNA microarrays *(3)*. The UniGene database creates nonredundant gene clusters based on GenBank sequences *(3)*. Clusters are built by sequence alignment and annotated relative to genes in the Entrez Gene database. Consequently, UniGene clusters can be thought of as collections of GenBank sequences that most likely describe the same gene.

The RefSeq database provides exemplary transcript and protein sequences based either on hand curation or based on information from a genome authority (e.g., the Jackson Labs) *(3,15)*. RefSeq accession numbers follow a PREFIX_NUMBER format (e.g., NM_123456, or NM_123456789). All curated RefSeq transcript accessions are prefixed by an NM, whereas XM prefixes represent accessions that have been generated by automated methods. Some of the NM transcript accessions are generated by automated methods, but are mature, and have undergone some level of review. RefSeq records

Table 1
RefSeq Status Codes and Their Level of Annotation*

RefSeq status code	Level of annotation
Genome annotation	Records that are aligned to the annotated genome
Inferred	Predicted to exist based on genome analysis, but no known mRNA/EST exists within GenBank
Model	Predicted based on computational gene prediction methods; a transcript sequence may or may not exist within GenBank
Predicted	Sequences from genes of unknown function
Provisional	Sequences represent genes with known functions, however they have not been verified by NCBI personnel
Validated	Provisional sequences that have undergone a preliminary review by NCBI personnel
Reviewed	Validated sequences that represent genes of known function that have been verified by NCBI personnel

*See http://www.ncbi.nlm.nih.gov/RefSeq/key.html#status.

also contain one of seven status codes, illustrating the state of maturity of the annotation (**Table 1**).

4. Annotation Databases

Annotation databases provide functional information for genes and may also catalogue the structure of the gene. They serve as an initial point for data interpretation of microarray data and hypothesis generation.

Entrez Gene is a part of NCBI's Entrez suite of bioinformatics tools. It provides information on genes that have a RefSeq or have been annotated by a genome annotation authority (e.g., Jackson Labs for mice) for several toxicology relevant species, including human, mouse, rat, and dog (*5*). Consequently, entries within Entrez Gene may have an associated NM (mature) or the XM (nonreviewed) RefSeq, or may not have an exemplary RefSeq sequence associated with it.

Entrez Gene serves as a focal point for the integration of gene annotation data from many sources, including databases outside NCBI. Some data integration is achieved through hyperlinks to the appropriate database entries, and others are catalogued on the detail page for that gene. **Table 2** lists several of the annotation categories and their sources. The most basic form of gene annotation is the gene name and the abbreviation, which can be used to initiate the functional annotation of a gene through literature

Table 2
Entrez Gene Annotation Categories and Sources*

Annotation categories	Source
Gene names and abbreviations/symbols	Publications and genome authorities
RefSeq sequence	RefSeq database
Genome position and gene structures	Genome databases
Gene function	Gene Ontology (GO) database, Gene References into Function (GeneRIF)
Expression data	Gene Expression Omnibus (GEO), EST tissue expression from GenBank

*Adapted from Maglott, D., Ostell, J., Pruitt, K. D., and Tatusova, T. (2005) Entrez Gene: gene-centered information at NCBI. *Nucleic Acids Res* **33** (Database Issue), D54–58.

searches. Entrez Gene also integrates data from the RefSeq, Gene Ontology (GO), Gene Expression Omnibus (GEO), Gene References into Function (GeneRIF), and GenBank databases. The RefSeq sequences, both mRNA and protein, facilitate sequence-based searching, such as identifying homologous genes, or identifying putative function based on protein domains. The GO database catalogues genes by their molecular function, cellular location, and biological process. Information regarding the tissue expression of genes can be obtained from the GenBank database, where the tissue localization for an EST is recorded, as well as the GEO-NCBI's gene expression repository *(3)*. GeneRIFs provide curated functional data and literature references, although it may not be the most up-to-date functional annotation available in the literature. Investigators are encouraged to facilitate GeneRIF updates by submitting suggestions directly to the NCBI through their update form: http://www.ncbi.nlm.nih.gov/RefSeq/update.cgi.

For human studies, the Online Mendelian Inheritance in Man (OMIM) database, the online version of the Mendelian Inheritance in Man *(16)*, provides linkages between human genes and diseases *(3,17)*. Output pages from the Entrez Gene provide links to OMIM, which is also searchable through the NCBI Entrez system. For many of the diseases within OMIM, a synopsis of the clinical presentation is provided in addition to links to the genes associated with the disease. PubMed citations are also made available through OMIM, with hyperlinks to the PubMed database entries. OMIM also contains information on known allelic variants and some polymorphisms *(17)*.

The Gene Ontology (GO; http://www.geneontology.org) *(6)* database is another source of gene functional annotative information. The database consists of an ontology (i.e., a catalogue of existents/ideas/concepts and their interrelationships; Ref. *18*) where terms exist within a directed acyclic graph (DAG;

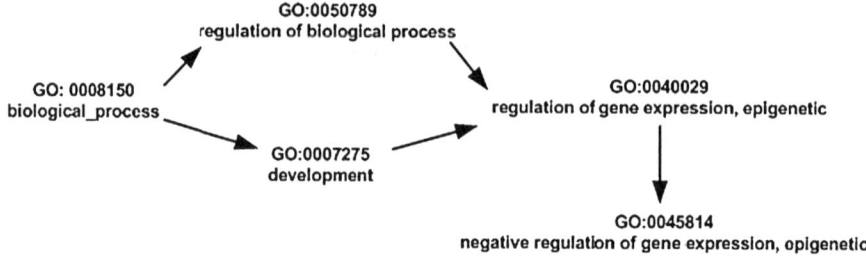

Fig. 3. Example of a Gene Ontology (GO) directed acyclic graph (DAG). This DAG shows two paths to reach the same GO entry, GO:0045814. It is important to note that the DAG travels from the most general case and becomes more specific with entries that are farther down the DAG.

Fig. 3). DAGs are graphical structures that cannot exist as loops, thus, a child node (i.e., an object or concept) may not also serve as its own predecessor (i.e., parent, grandparent, great-grandparent, etc.). Any child node within a DAG may have any number of parents and any number of paths to get to the child. For example, **Fig. 3** shows two paths leading to the same child, GO:0045814 (negative regulation of gene expression, epigenetic). The DAG illustrates that epigenetic negative regulation of gene expression is both a regulation process and critical in development. GO entries that exist at the same level relative to the root, or starting node, do not necessarily reflect the same level of specificity. The level of specificity afforded must be taken on a per DAG basis and not relative to the other DAGs. Thus, a fourth order node (a node that is four levels below the root node) in one DAG has no specificity relationship with regards to a sixth order node in a different DAG. At each mode within the GO there may exist a list of genes. As the annotation for a gene improves, it may change node associations. For example, if gene X were previously GO:0040029 (regulation of gene expression, epigenetic), and new experimental data suggested gene X was a negative regulator of gene expression through an epigenetic mechanism, it would be reassigned to GO:0045814 (negative regulation of gene expression, epigenetic).

The GO Consortium maintains the mappings between genes and the GO terms. It is important to note that each gene may have multiple associated GO terms and that the assignment of a GO number has no other significance other than being a unique identifier.

5. Protein-Level Databases

In many instances, the gene annotation databases mentioned above provide hyperlinks to protein annotation databases to identify the proteins encoded by the genes of interest. Recently, several protein-level databases were merged into one primary protein resource, the Universal Protein Resource (UniProt). UniProt combines the Swiss-Prot, TrEBML, and PIR-PSD databases into one resource, consisting of three related databases: *(1)* the UniProt Archive, *(2)* the UniProt Knowledgebase, and *(3)* the UniRef database.

The UniProt Archive (UniParc) is a database of nonredundant protein sequences obtained from *(1)* translation of sequences within the gene sequence level databases (e.g., GenBank), *(2)* RefSeq, *(3)* FlyBase, *(4)* WormBase, *(5)* Ensembl, *(6)* the International Protein Index, *(7)* patent applications, and *(8)* the Protein Data Bank *(19)*. The UniProt Knowledgebase (UniProt) provides functional annotation of the sequences within the UniParc. Examples of the annotation include the protein name, listing of protein domains and families from the InterPro database, containing protein family, domain, and functional information (http://www.ebi.ac.uk/interpro), *(20)*, Enzyme Commission identifier, and Gene Ontology identifiers. Proteins represented within the UniParc and UniProt Knowledgebase are then gathered automatically to create the UniProt reference database (UniRef), a database of reference, exemplary sequences based on sequence identity. Three different versions of the UniRef database exist (i.e., UniRef100, UniRef90, and UniRef50), where the number denotes the percent identity required for sequences to be merged, from across all species represented in the parent databases, into a single reference protein sequence. Thus, UniRef50 requires only 50% identity for proteins to be merged. UniRef50 and 90 provide faster sequence searches for identifying probable protein domains and functions by decreasing the size of the search space.

The RefSeq database also contains reference protein sequences, similar in concept to the reference mRNA sequences. These are available through the Entrez Gene system when querying for a gene. For more information on RefSeq, see **Section 3**.

6. Protein Interaction Databases

Protein interaction databases such as the Biomolecular Interaction Network Data (BIND) database, the Database of Interacting Proteins (DIP), the Molecular Interaction database (MINT), and the IntAct database provide information on the interaction of proteins with other proteins, genes, and small molecules. Both the BIND *(21)* and DIP *(8)* manage data from protein interaction experiments, including yeast-two-hybrid and co-immunoprecipitation experiments. This data is submitted to the databases either directly or as a result of database curators scouring the literature. The data is provided to the public through querying of the Web sites or in interaction files available in the Protein Standards Initiative (PSI) Molecular Interaction (PSI-MI) XML format.

Visualization of these data sets is made possible through tools such as Osprey *(22)* and Cytoscape *(23)*, which generate protein interaction networks based on input data from protein interaction databases or from other sources. Cytoscape has the additional functionality of allowing the overlay of gene expression data on the protein interaction map *(23)*. These visualization tools provide initial support in the elucidation of pathways that may be altered after treatment, facilitating the generation of new hypotheses and the identification of biomarkers of exposure and toxicity.

7. Microarray Databases

Microarray databases ensure data are being properly managed, support analysis, archive data for long-term use, and facilitate sharing with collaborators or deposition in public repositories. The Minimum Information About a Microarray Experiment (MIAME) standards provide guidance on the types of information that must be captured and reported in support of a microarray study in order to ensure independent investigators can replicate and properly interpret the data *(24)*. This includes information regarding the clones, genes, protocols, and samples associated with the study. Several journals require microarray submissions to adhere to the MIAME standard, and the MGED (Microarray Gene Expression Data) Society is encouraging journals to require that microarray data sets, in support of published articles, also be submitted to repositories as a condition of publication, similar to requirements that novel sequences be submitted to GenBank prior to publication *(25,26)*. Submission of microarray data sets to the NCBI Gene Expression Omnibus (GEO) *(12)* or the ArrayExpress *(10,11)* at the European Bioinformatics Institute (EBI) fulfills this requirement. Recently, more specialized repository efforts have been undertaken, such as the Chemical Effects in Biological Systems (CEBS) Knowledgebase *(9,27)*, which will catalogue gene expression data from chemical exposures with the associated pathology and toxicology data.

With the emergence of more pharmacology and toxicology domain specific data management systems, the International Life Sciences Institute (ILSI) Health and Environmental Sciences Institute (HESI) Technical Committee on the Application of Genomics to Mechanism-Based Risk Assessment, in cooperation with the MGED Society, began work on a toxicology-specific MIAME standard (MIAME/Tox) *(28)*. MIAME/Tox is expected to further specify the minimum information required to replicate a toxicogenomics experiment, which will also serve to facilitate data sharing among the toxicogenomics community. Moreover, it is expected that these databases will be extended to include the management of complementary proteomic and metabolomic data as well as other toxicology relevant data such as chemical/drug structure information, adsorption, distribution, metabolism, and excretion.

8. Conclusion

The use of genomic technologies in the mechanistic understanding of drug and chemical effects in biological tissues requires effective gene annotation. Several annotation sources exist; however, no database captures all of the data, making toxicogenomic data interpretation and network development difficult. For example, information concerning the function of a gene exists within Entrez Gene, however, protein family and structure information exist within the UniProt, and protein interaction data exist within databases such as BIND, DIP, and MINT. Ideally, the integration of data from these disparate sources into a single database would allow a more comprehensive interpretation of the available data. Moreover, a centralized comprehensive knowledgebase would also facilitate the identification of mechanistically based biomarkers for human toxicity and the development of computational models with greater predictive power, which could be used to support and improve quantitative risk assessments.

References

1. Clamp, M., Andrews, D., Barker, D., Bevan, P., Cameron, G., Chen, Y., et al. (2003) Ensembl 2002: accommodating comparative genomics. *Nucleic Acids Res.* **31**, 38–42.
2. Hubbard, T., Andrews, D., Caccamo, M., Cameron, G., Chen, Y., Clamp, M., et al. (2005) Ensembl 2005. *Nucleic Acids Res.* **33**(Database Issue), D447–453.
3. Wheeler, D. L., Church, D. M., Edgar, R., Federhen, S., Helmberg, W., Madden, T. L., et al. (2004) Database resources of the National Center for Biotechnology Information: update. *Nucleic Acids Res.* **32**, D35–40.
4. Karolchik, D., Baertsch, R., Diekhans, M., Furey, T. S., Hinrichs, A., Lu, Y. T., et al. (2003) The UCSC Genome Browser Database. *Nucleic Acids Res.* **31**, 51–54.
5. Maglott, D., Ostell, J., Pruitt, K. D., and Tatusova, T. (2005) Entrez Gene: gene-centered information at NCBI. *Nucleic Acids Res* **33**(Database Issue), D54–58.
6. Harris, M. A., Clark, J., Ireland, A., Lomax, J., Ashburner, M., Foulger, R., et al. (2004) The Gene Ontology (GO) database and informatics resource. *Nucleic Acids Res.* **32**, D258–261.
7. Bader, G. D. and Hogue, C. W. (2000) BIND—a data specification for storing and describing biomolecular interactions, molecular complexes and pathways. *Bioinformatics* **16**, 465–477.
8. Xenarios, I., Rice, D. W., Salwinski, L., Baron, M. K., Marcotte, E. M., and Eisenberg, D. (2000) DIP: the database of interacting proteins. *Nucleic Acids Res.* **28**, 289–291.

9. Waters, M., Boorman, G., Bushel, P., Cunningham, M., Irwin, R., Merrick, A., et al. (2003) Systems toxicology and the Chemical Effects in Biological Systems (CEBS) knowledge base. *EHP Toxicogenomics* **111**, 15–28.

10. Brazma, A., Parkinson, H., Sarkans, U., Shojatalab, M., Vilo, J., Abeygunawardena, N., et al. (2003) ArrayExpress—a public repository for microarray gene expression data at the EBI. *Nucleic Acids Res.* **31**, 68–71.

11. Rocca-Serra, P., Brazma, A., Parkinson, H., Sarkans, U., Shojatalab, M., Contrino, S., et al. (2003) ArrayExpress: a public database of gene expression data at EBI. *C. R. Biol.* **326**, 1075–1078.

12. Edgar, R., Domrachev, M., and Lash, A. E. (2002) Gene Expression Omnibus: NCBI gene expression and hybridization array data repository. *Nucleic Acids Res.* **30**, 207–210.

13. Curwen, V., Eyras, E., Andrews, T. D., Clarke, L., Mongin, E., Searle, S. M., and Clamp, M. (2004) The Ensembl automatic gene annotation system. *Genome Res.* **14**, 942–950.

14. Rouchka, E. C., Gish, W., and States, D. J. (2002) Comparison of whole genome assemblies of the human genome. *Nucleic Acids Res.* **30**, 5004–5014.

15. Pruitt, K. D. and Maglott, D. R. (2001) RefSeq and LocusLink: NCBI gene-centered resources. *Nucleic Acids Res.* **29**, 137–140.

16. McKusick, V. A. (1998) *Mendelian Inheritance in Man. A Catalog of Human Genes and Genetic Disorders.* 12th ed. Johns Hopkins University Press, Baltimore.

17. Hamosh, A., Scott, A. F., Amberger, J., Bocchini, C., Valle, D., and McKusick, V. A. (2002) Online Mendelian Inheritance in Man (OMIM), a knowledgebase of human genes and genetic disorders. *Nucleic Acids Res.* **30**, 52–55.

18. Cox, C. (1999) *Nietzsche: Naturalism and Interpretation.* University of California Press, Berkeley.

19. Bairoch, A., Apweiler, R., Wu, C. H., Barker, W. C., Boeckmann, B., Ferro, S., et al. (2005) The Universal Protein Resource (UniProt). *Nucleic Acids Res* **33** (Database Issue), D154–159.

20. Mulder, N. J., Apweiler, R., Attwood, T. K., Bairoch, A., Barrell, D., Bateman, A., et al. (2003) The InterPro Database, 2003 brings increased coverage and new features. *Nucleic Acids Res.* **31**, 315–318.

21. Alfarano, C., Andrade, C. E., Anthony, K., Bahroos, N., Bajec, M., Bantoft, K., et al. (2005) The Biomolecular Interaction Network Database and related tools 2005 update. *Nucleic Acids Res.* **33**(Database Issue), D418–424.

22. Breitkreutz, B. J., Stark, C., and Tyers, M. (2003) Osprey: a network visualization system. *Genome Biol.* **4**, R22.

23. Shannon, P., Markiel, A., Ozier, O., Baliga, N. S., Wang, J. T., Ramage, D., et al. (2003) Cytoscape: a software environment for integrated models of biomolecular interaction networks. *Genome Res.* **13**, 2498–2504.

24. Brazma, A., Hingamp, P., Quackenbush, J., Sherlock, G., Spellman, P., Stoeckert, C., et al. (2001) Minimum information about a microarray experiment (MIAME)—toward standards for microarray data. *Nat. Genet.* **29**, 365–371.

25. Ball, C. A., Brazma, A., Causton, H., Chervitz, S., Edgar, R., Hingamp, P., et al. (2004) Submission of microarray data to public repositories. *PLoS Biol.* **2**, E317.
26. Ball, C. A., Sherlock, G., and Brazma, A. (2004) Funding high-throughput data sharing. *Nat. Biotechnol.* **22**, 1179–1183.
27. Waters, M. D., Olden, K., and Tennant, R. W. (2003) Toxicogenomic approach for assessing toxicant-related disease. *Mutat. Res.* **544**, 415–424.
28. Mattes, W. B., Pettit, S. D., Sansone, S. A., Bushel, P. R., and Waters, M. D. (2004) Database development in toxicogenomics: issues and efforts. *Environ. Health Perspect.* **112**, 495–505.

8

Microarray Probe Mapping and Annotation in Cross-Species Comparative Toxicogenomics

John N. Calley, William B. Mattes, and Timothy P. Ryan

Summary

Genomics-based tools, such as microarrays, do appear to offer promise in evaluating the relevance of one species to another in terms of molecular and cellular response to a given treatment. However, to fulfill this promise the individual end points (i.e., the genes, proteins, or mRNAs) measured in one species must be mapped to corresponding end points in another species. Several approaches, along with their strengths and weaknesses, are described in this chapter. A sequential approach is described that first makes use of a "Genome To Genome Through Orthology" method, where probe sequences for a given species are mapped into full-length sequences for that species, associated with the locus for those sequences and then into a second species by consulting orthology resources. The second step supplements these results by mapping the probe sequences for the given species into the best matching transcript from any organism, which then are mapped into the appropriate native locus and finally into the second species via an orthology resource. The results of this method are given for an experiment comparing the transcriptional response of canine liver to phenobarbital with that of rat liver.

Key Words: annotation; bioinformatics; cross-species; genomics; microarray; mRNA.

From: *Methods in Molecular Biology, vol. 460: Essential Concepts in Toxicogenomics*
Edited by: D. L. Mendrick and W. B. Mattes © Humana Press, Totowa, NJ

1. Introduction

1.1. Why Perform Cross-Species Comparisons of Microarray Experiments?

Basic research into the molecular underpinnings of biological responses obviously requires the choice of an experimental system. Whether that system is a cell-free preparation, an isolated tissue, or a whole organism, the system ultimately owes its basic characteristics to the genetic program of the species from which it was derived. For example, Birch and Schreiber *(1)* used a rat liver cell free system to investigate gene expression responses during acute inflammation, and whereas the simplest assumption is that such responses will be common in other (mammalian) species, that assumption is a hypothesis until tested *(2)*.

The question of whether the results seen in one species are relevant to what may be expected in another is particularly critical in the field of toxicology. As noted in the Society of Toxicology's Animals in Research Public Policy Statement, "research involving laboratory animals is necessary to ensure and enhance human and animal health and protection of the environment." Thus, toxicity results seen in an animal model are extrapolated to other species, including man. Given the inherent uncertainty in such extrapolations, programs designed to assess the human safety of a potential medicine mandate that such testing is conducted in more than one species, with the hope that one of these may reflect the human response. Indeed, the International Conference on Harmonisation (ICH) guideline for Non-Clinical Safety Studies for the Conduct of Human Clinical Trials for Pharmaceuticals M3 states that "... in principle, the duration of the animal toxicity studies conducted in two mammalian species (one non-rodent) should be equal to or exceed the duration of the human clinical trials up to the maximum recommended duration of the repeated dose toxicity studies." Of course, when results in one test species are concordant with those in another, the confident assumption is that the results reflect a biological response that will be seen in several species, including humans. On the other hand, when the results in two test species are discordant, the question remains as to which biological response may be expected in humans (or a third species).

Genomic information, both in terms of the primary genome sequence and in terms of global views of gene expression, does appear to offer promise in evaluating the relevance of one species to another in terms of molecular and cellular response to a given treatment. Thus, mRNA profiling with microarrays allows a global view of toxicant-induced transcriptome alterations in various cell types, tissues, and species and allows an experimental view of similarities and differences in signaling and response pathways with previously unattainable granularity. The promise is that such tools will allow a truly molecular assessment of toxic responses, allowing better assessment of the relevance of various animal models to one another and to man.

2. Mapping of Microarray Probes Across Species

A necessary precondition for any correlation of microarray results across species is a mapping of the reagents (probes or probesets) into a common set of identifiers. There are many ways that this mapping can be performed, and the common identifiers can be at several different levels of abstraction. These levels include gene function, locus, sequence, and probe level analysis and are outlined in **Fig. 1** and described in more detail below.

2.1. Direct Chip to Chip Mapping

The lowest level of abstraction, when comparing microarrays from two different organisms, is direct comparison of the sequences of probes on one microarray with those on the other. This is shown in gray in **Fig. 1** to indicate that this is rarely advisable. Although it may be attractive from the standpoint of computational tractability, the trade-offs in accuracy are high. In particular,

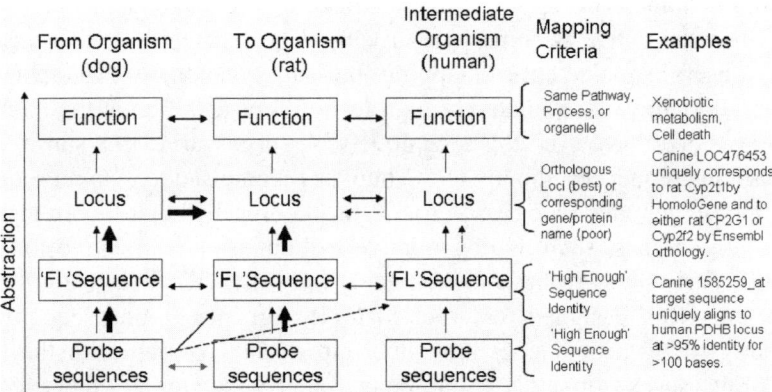

Fig. 1. Mappings at the sequence level are the simplest but require high levels of sequence conservation between the organisms and well-characterized genomes. Mapping directly between probe sequences is almost never advisable. "FL" sequence can be cDNA sequence and/or genomic sequence for a locus. Mapping at the locus level can help when loci are poorly characterized and can also help alert one to problematic orthology. Mapping at the functional level is the most complex but relieves one of the need to sample exactly the same loci in both experiments. The path marked in bold lines is the "genome to genome" method primarily used here. Indicated in dashed lines is the path used for supplementary annotations through human cDNA comparisons.

it is rare for probe sequences on two independently designed microarrays to be derived from the same portion of a message. Because of inevitable differences in chip design, some probes will be derived from 3′ UTR (Untranslated Region) sequence, some from coding sequence, some from 5′ UTR, and some may even be from alternative exons or alternative UTRs. For these reasons, even when both chips contain sequences from the same organism, directly matching probe sequences is a poor way of correlating array data. When comparing microarray data between two different organisms, the problem of false negatives becomes very evident because the probes may be in relatively unconserved regions of the sequence. Additionally, it is inevitable that mistakes will be made in assessing orthology inappropriately as a result of lack of information about genes that are not on the chip. Just because two gene sequences on the two different chips are similar does not mean that these are the most similar possible sequences over the two complete genomes.

2.2. Mapping Two Chips to a Single Genome

When two organisms to be compared have incomplete or poorly annotated genomic sequence, it may be essential to associate probes with loci in a better-characterized genome. This can be the better characterized of the two genomes or a third genome. This approach is illustrated by the diagonal arrow in **Fig. 1**. Even when the native genome is relatively well characterized, this technique can be a useful supplement to mapping into native genomes. This approach is only appropriate when a relatively high level of conservation at the nucleotide level exists. We have used this technique with success to map a subset of dog and macaque sequences directly to the human genome and to map rat sequences to the mouse genome. One reason to use this approach is that loci in less well-annotated genomes, such as dog relative to human or rat relative to mouse, are particularly likely to have incompletely known UTRs. Because microarray probes are often designed to target 3′ UTR sequence, probe sequences are frequently found near one or more loci but do not overlap with the known range of messages from a particular locus. If a probe shares sufficient sequence identity at the nucleotide level to pass a reasonably stringent identity filter when compared with the organism with a better characterized genome, and this relationship is unique within the target genome, our experience has shown that automatic identifications using this method are usually confirmed by more labor-intensive methods.

There are several alternative approaches to solving this problem that may be appropriate. First, associating a probe (or the sequence from which it is derived) with any locus that is "close enough" in the genome (and on the correct strand!). Second, using a transcript clustering resource like UniGene to associate available ESTs with locus identifiers. Finally, *de novo* EST clustering and assembly followed by association of the extended sequence with a locus. Each of these approaches has its place. Experience has shown that the first two methods have relatively high error rates, limiting their application to speculative associations. The last method, *de novo* EST clustering, although very productive, and allowing the use of nonpublic sequence resources, is highly computationally intensive and also error prone. Therefore, the direct cross-species method is generally more straightforward and in our hands performs quite well when appropriately applied. When time and resources permit, it is best to use several of these methods and combine their results.

Table 1 and **Table 2** demonstrate four cases where mapping canine probesets to human sequences has allowed at least tentative identifications or has strengthened our confidence in weak associations. This method is illustrated by dotted arrows on **Fig. 1**. Both GL-Cf-261_at and GL-Cf-4307_at, for example, could not be mapped directly to the canine genome but were associated with human transcripts and then, through orthology, to rat P450 genes. In both cases, the exact P450 orthologue is ambiguous, but at least some information was gained. Similarly, GL-Cf-6542_at and GL-Cf-6563_at were both mapped to the canine genome but could only be associated with a canine locus via the weak "close enough" criterion. The fact that matches to human transcripts yielded the same identifications significantly increases their reliability.

2.3. Mapping Between Orthologous Native Loci

When complete and reasonably well-characterized genomic information is available for both species to be compared, association of microarray probe sequences with the appropriate loci in the native genome, followed by associating orthologous loci between the two organisms, is preferred. To do this, first associate the microarray probe sequences with either full-length cDNA sequences or with annotated genomic sequence (or both) and then identify the locus associated with the matched sequence. In Entrez Gene, for example, one can identify matching RefSeq transcripts and then consult gene2refseq to look up the corresponding Entrez Gene locus ID. Alternatively for genomic matches, one can consult the seqgene.md file from NCBI to associate a match to a particular genomic stretch with an exon from an Entrez Gene locus. For Ensembl, we download the annotated GenBank formatted genomes to drive the corresponding associations. Because of the different gene annotation methods used by the NCBI and Ensembl, results are not equivalent, and we have found it informative to use both when possible. This is the "genome to genome through orthology" method we have primarily relied on for the experimental analysis done here and is indicated in bold on **Fig. 1**.

Table 1

Probeset	% present vehicle	% present phenobarbital	FC signed magnitude (vehicle vs. phenobarbital)	t-test p value (vehicle vs. phenobarbital)	Native locus match	Native locus symbol
GL-Cf-10675_at	33.33	33.33	1.16	.48	E:3ge, N:3ge	LOC483120
GL-Cf-11066_at	100	100	1.03	.87	E:g0, N:g0	CYP19
GL-Cf-11386_at	100	100	-1.06	.78	E:g0	Cyp4a10
GL-Cf-11716_at	100	100	**1.92**	**1.82E-02**	E:g1,tr, N:g1,tr	LOC475225
GL-Cf-1188_at	100	100	-1.29	8.40E-02	N:3ge, E:3ge	CYP4A39
GL-Cf-13357_i_at	100	100	-1.19	**3.33E-02**	N:3ge	LOC607548
GL-Cf-136_i_at	100	100	**1.83**	.12	N:g1,tr	CYP2D15 CYP2D
GL-Cf-14115_at	100	100	-1.17	.42	N:3ge	LOC607548
GL-Cf-1422_at	0	0			E:g1,tr, N:g1,tr	LOC478365
GL-Cf-1594_at	100	100	-1.5	.27	N:g1,tr, E:tr	CYP2A13 LOC487121
GL-Cf-179_s_at	100	100	1.37	**1.88E-03**	E:3ge,N:g1	CYP2B11
GL-Cf-1940_at	100	100	**1.87**	**2.51E-02**	N:g1,tr, E:tr	CYP3A26
GL-Cf-1989_at	100	100	-1.04	.48	N:3ge, E:g0	CYP4A37
GL-Cf-261_at	33.33	33.33	**2.7**	.73	N.A.	
GL-Cf-3134_at	100	100	1.26	9.72E-02	E:3ge, N:g1,tr	LOC610195
GL-Cf-339_s_at	100	100	1.02	.79	N:g1,tr, E:tr	CYP2C21
GL-Cf-407_r_at	100	100	-1.05	.26	E:g1,tr, N:g1,tr	LOC479740
GL-Cf-4307_at	100	100	1.3	.21	N.A.	
GL-Cf-4324_at	100	100	**2.33**	**3.69E-03**	E:g1,tr, N:g1,tr	LOC475225
GL-Cf-4519_at	0	0			E:g1,tr, N:g1,tr	LOC477807

GL-Cf-4541_at	100	−1.47	.32	E:g1,tr, N:g1,tr	LOC479286
GL-Cf-46_at	0			E:g1	CYP1A1
GL-Cf-521_at	0			N:g1, E:g1,tr	CYP21A
GL-Cf-5294_at	0			E:g0, N:g0	CYP19
GL-Cf-5559_r_at	100	−1.17	.1	E:g1,tr, N:g1,tr	CYP2E1
GL-Cf-6542_at	100	1.01	.97	E:3ge, N:3ge	LOC483038
GL-Cf-6563_at	100	−1.35	.12	E:3ge, N:3ge	LOC483038
GL-Cf-71_s_at	100	−1.2	.22	E:g1,tr, N:g1,tr	CYP2E1
GL-Cf-8060_at	100	−1.17	.38	N:3ge	LOC607548
GL-Cf-8700_at	0			N:g1, E:g1,tr	CYP19
GL-Cf-9219_at	100	**1.95**	**2.95E-02**	N:g1,tr, E:g0	Cyp3a13 Cyp3a18 LOC489851
GL-Cf-958_i_at	100	**2.74**	**4.97E-02**	N:g1	CYP2D15 CYP2D

All Affymetrix probesets for cytochrome P450 are listed with the exception of those with "Absent" calls in all samples. *Native locus match* describes the methods used to map the target sequence for the indicated probeset to the canine genome. "E:" refers to matches made via the Ensembl annotated canine genome. "N:" refers to matches made via the NCBI annotated canine genome. "3ge" means that a genomic match was found within 5 kb of the annotated 3′ end of the gene. "g0" means that a match was found within an intron of the gene. "g1" means that a match was found that overlapped with an annotated exon. "tr" means that a match was found to a predicted or known transcript sequence from the indicated gene. Bolded text denotes probe sets that met the statistical criteria noted in the text (mean group values ≥1.8-fold difference between control and phenobarbital groups with *t*-test *p* values <.05 and a 100% present call in either the control or phenobarbital group)

Table 2

Probeset	Native locus symbol	Non-native match	Orthology	Reliability	Gene symbol	Associated organism	RAE230_2 probesets
GL-Cf-10675_at	LOC483120	N.D.	H:unique, E:UBRH	Low	Cyp26b1	Rat	1384392_at (High), 1376667_at (High)
GL-Cf-11066_at	CYP19	N.D.	H:unique, E:UBRH	Low	Cyp19a1	Rat	1369444_at (High), 1379643_at (Medium)
GL-Cf-11386_at	Cyp4a10	N.D.	E:UBRH	Low	Cyp4a10	Rat	
GL-Cf-11716_at	LOC475225	N.D.	H:unique, E:UBRH	High	Cyp51	Rat	1367979_s_at (High), 1387020_at (Medium)
GL-Cf-1188_at	CYP4A39	N.D.	H:unique, E:UBRH	Ambiguous	Multiple (Cyp4a10; Cyp4a12)	Rat	
GL-Cf-13357_i_at	LOC607548	N.D.	H:unique	Low	Cyp4v3	Mouse	
GL-Cf-136_i_at	CYP2D15 CYP2D	N.D.	H:unique	Medium	Cyp2d22	Rat	1387913_at (High), 1370329_at (Medium)
GL-Cf-14115_at	LOC607548	N.D.	H:unique	Low	Cyp4v3	Mouse	
GL-Cf-1422_at	LOC478365	N.D.	H:unique, E:UBRH	High	Cyp11a1	Rat	1368468_at (High)
GL-Cf-1594_at	CYP2A13 LOC487121	N.D.	H:multiple, E:UBRH	Ambiguous	Multiple (Cyp2a4; CYP2A6; CYP2A7)	Human, mouse	
GL-Cf-179_s_at	CYP2B11	N.D.	H:unique, E:UBRH	Medium	Cyp2b10	Mouse	1371076_at (Low)
GL-Cf-1940_at	CYP3A26	N.D.	H:multiple, E:RHS	Ambiguous	Multiple (Cyp3a44; Cyp3a11; Cyp3a13; Cyp3a41)	Rat, mouse	

Probe	Gene		Annotation	Quality	Ortholog	Species	Affymetrix probes
GL-Cf-1989_at	CYP4A37	N.D.	H:multiple, E:UBRH	Low	Cyp4a10	Rat	
GL-Cf-261_at	N.A.	Hs:tr	H:unique	Ambiguous	Multiple (Cyp2c37, Cyp2c)	Rat	
GL-Cf-3134_at	LOC610195	N.D.	H:unique, E:UBRH	Medium	Cyp2j4	Rat	1387296_at (High), 1396327_at (Low)
GL-Cf-339_s_at	CYP2C21	N.D.	H:unique, E:UBRH	Ambiguous	Multiple (Cyp2c70; Cyp2c39)	Rat, mouse	
GL-Cf-407_r_at	LOC479740	N.D.	H:unique, E:UBRH	High	Cyp3a13	Rat	1370387_at (High)
GL-Cf-4307_at	N.A.	Hs:tr	H:unique, E:RHS	Ambiguous	Multiple (Cyp2c6, Cyp2, Cyp2c70)	Rat	
GL-Cf-4324_at	LOC475225	N.D.	H:unique, E:UBRH	High	Cyp51	Rat	1367979_s_at (High), 1387020_at (Medium)
GL-Cf-4519_at	LOC477807	N.D.	H:unique, E:UBRH	High	Cyp17a1	Rat	1387123_at (High)
GL-Cf-4541_at	LOC479286	N.D.	H:unique, E:UBRH	High	Cyp1a2	Rat	1387243_at (High)
GL-Cf-46_at	CYP1A1	N.D.	E:UBRH	Medium	Cyp1a1	Rat	1370269_at (Medium)
GL-Cf-521_at	CYP21A	N.D.	H:unique, E:UBRH	Medium	Cyp21a1	Rat	1369264_at (High)
GL-Cf-5294_at	CYP19	N.D.	H:unique, E:UBRH	Low	Cyp19a1	Rat	1369444_at (High), 1379643_at (Medium)
GL-Cf-5559_r_at	CYP2E1	N.D.	H:unique, E:UBRH	High	Cyp2e1	Rat	1367871_at (High)
GL-Cf-6542_at	LOC483038	Hs:tr	H:unique, E:UBRH	Medium	Cyp1b1	Rat	1368990_at (High)

Table 2
(Continued)

Probeset	Native locus symbol	Non-native match	Orthology	Reliability	Gene symbol	Associated organism	RAE230_2 probesets
GL-Cf-6563_at	LOC483038	Hs:tr	H:unique, E:UBRH	Medium	Cyp1b1	Rat	1368990_at (High)
GL-Cf-71_s_at	CYP2E1	N.D.	H:unique, E:UBRH	High	Cyp2e1	Rat	1367871_at (High)
GL-Cf-8060_at	LOC607548	N.D.	H:unique	Low	Cyp4v3	Mouse	
GL-Cf-8700_at	CYP19	N.D.	H:unique, E:UBRH	Medium	Cyp19a1	Rat	1369444_at (High), 1379643_at (Medium)
GL-Cf-9219_at	Cyp3a13 Cyp3a18 LOC489851	N.D.	H:unique, E:UBRH	Ambiguous	Multiple (Cyp3a13; Cyp3a18)	Rat	
GL-Cf-958_i_at	CYP2D15 CYP2D	N.D.	H:unique	Medium	Cyp2d22	Rat	1387913_at (High), 1370329_at (Medium)

In selected cases, matches with human transcripts were also checked (at reduced stringency); these are recorded in the *Non-native match* column. *Orthology* describes the method used to move from the identified (usually) canine gene to the (usually) rat gene used to make the association with the rat microarray. "H:" means the association was derived from HomoloGene. "Unique" means that the HomoloGene entry had only single entries for both the species involved. "Multiple" means that the HomoloGene entry had more than one match in one or both species. "E:" means that the association was derived from Ensembl's orthology calls. "UBRH" refers to Ensembl's Unique Best Reciprocal BLAST Hit criterion. "RHS" refers to Ensembl's synteny criterion. *Reliability* is a summary of the weight of the evidence that the association between the canine probe, the canine gene, and the rat gene is reliable. *Gene symbol* is the (usually) rat gene symbol. In a few cases, it was not possible (at least via automated methods) to associate directly with a rat gene. *Associated organism* is the name of the organism from which the gene symbol is derived. *RAE230_2 Probesets* is the names of probe sets on the Affymetrix RAE230_2 Rat array for the indicated genes. Reliabilities of each association are given after each probeset name.

Once probe sequences are associated with native loci, the next step is to determine if there are "corresponding genes" between organisms. To determine gene correspondence there are two basic approaches: gene name based and sequence based.

Gene name–based approaches exploit prior efforts of model organism research communities to standardize gene nomenclature. These efforts have progressed well, especially in mammals, but there is still a long way to go before this is a reliable method to compare genes between species. When investigating organisms with newer genomes such as the dog, this proves problematic. Even within an organism, there are many instances where the same gene name is used to refer to two different genes in different technical literatures. One example of this duplicate nomenclature is the name FAS, which refers either to tumor necrosis factor receptor 6 or to fatty acid synthase, depending on whether one derives the information from the apoptosis or the lipid metabolism literature. Even more significant than the threat of making inappropriate associations with this approach, however, is that correct associations will be missed. Making use of a set of gene synonyms from, for example, Entrez Gene helps with this problem significantly but also increases inappropriate associations. An attempt should be made to purge these lists of those names (like FAS) that have two or more distinct references within either organism. Comparison of the columns labeled "Native locus symbol" and "Gene symbol" in **Table 1** to **Table 4** illustrates this point.

A usually better, although more difficult, approach to identify corresponding genes is based on the evolutionary history of two genes. Here genes are differentiated based on whether they are *orthologues* or *paralogues*. The distinction between *orthologues* (split due to speciation) and *paralogues* (split due to gene duplication) was made by Walter Fitch *(3)* to clarify when the relationship between two gene sequences is informative for the relationship between two species. Orthologous genes are more likely to be functionally conserved than paralogous genes because paralogous genes can drift in function after they are duplicated *(4–6)*. Orthology, as defined above, is not, of course, a guarantee of shared function in an organismal context. For example, while the primary amino acid sequence of CYP1A1 is highly conserved across species, its substrate specificity can be different in different species *(7)*. A similar observation has been made for the sulfotransferases *(8)*. Nonetheless, in these cases the *general* function is shared across species.

There are many different resources that provide precomputed assessments of orthology. Sources that we have found to be valuable in this way include HomoloGene *(9)*, Ensembl *(10)*, KOG/COG *(9)*, MGI *(11)*, inParanoid *(12)*, OrthoMCL *(4)*, and BioBASE's Proteome *(13,14)*. These resources differ in their methods, in the organisms covered, and in the locus identifiers used.

Although arguably the least interesting, the simple issue of locus identification is often the biggest practical barrier to using a particular orthology resource. InParanoid's results, for example, are provided primarily in the form of Ensembl gene identifiers. Entrez Gene identifiers then, must be translated into Ensembl gene identifiers before inParanoid can be used. Although such translations exist, our experience is that these seemingly simple translations can be highly problematic in practice. Because of the theoretical and practical issues involved in mapping into orthologous loci, we highly recommend using two different methods so that their results can be compared. This approach was used in the analysis shown in **Table 1** to **Table 4**, where orthology results using Ensembl Gene IDs and Ensembl orthology are correlated with those obtained using Entrez Gene IDs and HomoloGene orthology.

2.4. Mapping at the Functional Level

All of the above methods require that corresponding individual genes be measurable in both experiments to be compared. When it is possible to associate nearly complete functional information with a locus, however, there is a very attractive alternative, or supplement, to the techniques described above. When sufficient functional information is available, the preferred means of comparison is at the level of the affected biological processes. For example, if both organisms are undergoing an inflammatory response, and there is a reasonable sampling of genes involved in this process from both organisms, the biological perturbations can be compared, even when the genes involved are not the same. For the two chips considered in the current example, 164 genes are classified as having a role in the inflammatory response on the Gene Logic canine chip and 349 on the RAE230_2 rat chip (data not shown). Of these, approximately 136 are common to both chips when all reliabilities are considered. When only those associations of high reliability are considered, the number in common drops to 38. We are much more likely to recognize the existence of an inflammatory response when we have a larger sample of potentially informative genes. We are far better off then using all the available genes on both chips to understand the biological responses within each organism and then to correlate these responses, rather than limiting ourselves to the inevitably smaller, less informative set of genes that are shared between the two experiments.

3. Detailed Association Methods

For illustrative purposes we conducted a sample cross-species association between rat and dog that is reported in **Table 3** and **Table 4**. We used the Gene Logic (Gaithersburg, MD) Canine microarray for dog and the standard

RAE230_2 microarray (from Affymetrix, Santa Clara, CA) for rat. We took two of the paths illustrated in **Fig. 1**. First, we took a path vertically where we started with canine probe target sequences, mapped them into full length (FL) sequences in dog, looked up the locus associated with these FL sequences, and then moved horizontally into rat by consulting orthology resources. This is the "genome to genome through orthology" method and is illustrated in **Fig. 1** by the bold arrows. Second, we supplemented the results of this by taking a route where we mapped the canine probe sequences into the best matching FL transcript from any organism and then mapped these FL sequences into the appropriate native locus and into rat via an orthology resource. This path is illustrated in **Fig. 1** by the dotted arrows. Each of these methods is described in detail below.

3.1. Mapping Between Orthologous Native Loci

We took the "target sequences" (also known as SIF sequences) corresponding with each of the probesets on the two chips and used BLAT ver. 26 *(15)* to identify all matches in both genomic sequence and RNA sequence for Canine Build 2 downloaded from NCBI and for Canine Build 1 from Ensembl. The use of an older build was necessary with Ensembl because at the time no Ensembl annotation was available for Build 2. Matches were recorded when at least 70% of the target sequence matched at the 90% level or better. Genomic matches were further characterized as to whether they overlapped with at least one known exon (g1), were within known gene bounds but not overlapping a known exon (g0), were within 5 kb of the 3′ end of a known gene and not within 10 kb of the 5′ end of a different known gene (3ge), matched a transcript sequence from a known gene (tr), and were on the correct strand to be assayed by the microarray. These matching genomic regions and transcript sequences were then mapped into known loci in the form of Entrez Gene IDs (for NCBI matches) or Ensembl Gene IDs. For NCBI genomic matches, this was done via the coordinates in the seq_gene.md file and for transcript matches via the gene2refseq file, both available from the NCBI ftp site. For Ensembl genomic matches, this was done using the GenBank formatted genome sequence (via ftp) and for transcript matches via a correspondence table generated with EnsMart *(10)*.

NCBI canine Entrez Gene IDs were then looked up in HomoloGene *(9)* and, when possible, converted into rat Entrez Gene IDs along with an indication as to whether the correspondence was one-to-one ("unique") or not ("multiple") as a quality metric. When no rat Entrez Gene ID was available in the HomoloGene entry, a mouse ID was used; if no mouse was available, a human entry was used.

Table 3

Probeset	% present vehicle	% present phenobarbital	FC signed magnitude (vehicle vs. phenobarbital)	t-test p value (vehicle vs. phenobarbital)	Native locus match	Native locus symbol
GL-Cf-10423_at	100	100	2.48	4.71E-03	E:3ge	Fads3
GL-Cf-1043_at	100	100	−2.54	2.60E-02	E:g1,tr,N:g1,tr	LOC489393
GL-Cf-10437_at	100	100	2.25	2.99E-02	E:g1,tr,N:g1,tr	LOC607852
GL-Cf-1049_at	100	100	2.36	2.65E-02	E:3ge,N:g1,tr	LOC479182
GL-Cf-1062_at	100	100	1.87	3.49E-02	E:g1,tr,N:3ge+	PPP2CA
GL-Cf-10667_at	66.67	100	2.38	3.96E-02	E:g1,tr,N:g1,tr	SERPINE1 PAI-1
GL-Cf-11061_at	100	100	−2.25	3.01E-03	E:3ge,N:3ge	LOC479363
GL-Cf-11636_at	100	100	2.79	1.11E-02	E:3ge,N:g1,tr	LOC610935
GL-Cf-11716_at	100	100	1.92	1.82E-02	E:g1,tr,N:g1,tr	LOC475225
GL-Cf-11757_r_at	100	100	2	5.17E-03		
GL-Cf-12118_at	100	100	1.81	2.55E-02		
GL-Cf-12207_at	100	100	2.28	6.16E-03	N:g1,tr	LOC607852
GL-Cf-12295_at	100	100	2.16	1.31E-04	E:3ge,N:3ge	LOC612549
GL-Cf-12678_at	33.33	100	2.41	1.73E-02	E:3ge,N:g1,tr	FBN1
GL-Cf-13327_at	100	100	−2.2	1.94E-02		
GL-Cf-1344_at	100	100	2.17	1.54E-02	E:g1,tr,N:g1,tr	LOC476600
GL-Cf-13448_at	100	100	−1.84	1.10E-02		
GL-Cf-13922_at	0	100	2	1.03E-04	E:g1,tr,N:g1,tr	LOC475926 LOC607806
GL-Cf-13983_at	66.67	100	1.94	1.28E-02	E:3ge,N:3ge	LOC612198
GL-Cf-14020_at	100	100	1.89	3.54E-02	E:3ge,N:g1,tr	LOC487139
GL-Cf-14153_at	100	66.67	−3.36	3.58E-02	E:g1,N:g1	IL1RN IL-1ra

Probe	%	%	Value	p-value	Mapping	Gene
GL-Cf-14195_at	100	100	−2.33	1.83E-02	E:g1,tr,N:g1,tr	LOC474666
GL-Cf-1575_at	100	100	1.84	2.85E-02	E:3ge,N:g1,tr	LOC475939
GL-Cf-1587_at	100	100	−1.9	6.64E-03	E:g1,tr,N:g1,tr	LOC478665
GL-Cf-1937_at	33.33	100	1.98	4.74E-02	E:g1	SEC22A
GL-Cf-1940_at	100	100	1.87	2.51E-02	E:tr,N:g1,tr	CYP3A26
GL-Cf-1979_at	100	100	3.04	1.47E-02	E:g1,tr,N:g1,tr	Fads3 LOC483793
GL-Cf-2178_at	33.33	100	2.24	4.47E-02	E:g1,tr,N:g1,tr	LOC479147
GL-Cf-2228_at	100	100	2.01	7.19E-03	E:g1,tr,N:g1,tr	LOC478664
GL-Cf-235_at	100	100	2.32	2.36E-02	E:g1,tr,N:g1,tr	DVS27
GL-Cf-274_at	0	100	2.92	1.15E-02	E:g1,tr,N:g1,tr	PLA2G7
GL-Cf-2756_at	100	100	−2.25	3.08E-02		
GL-Cf-3098_at	100	100	2.59	2.05E-02	E:g1,tr,N:g1,tr	LOC488291
GL-Cf-380_at	100	100	−1.86	1.05E-02	E:g1,tr,N:g1,tr	CXADR
GL-Cf-3826_at	100	100	−1.84	3.65E-02	E:g1,tr,N:g1,tr	LOC475175
GL-Cf-3844_at	100	100	7.64	1.85E-03	E:g1,tr,N:g1,tr	LOC474802
GL-Cf-4182_at	100	100	2.34	3.23E-02	E:g1,tr,N:g1,tr	LOC480220
GL-Cf-4202_at	100	100	2.09	2.29E-02	E:g1,tr,N:g1,tr	LOC483378
GL-Cf-4324_at	100	100	2.33	3.69E-03	E:g1,tr,N:g1,tr	LOC475225
GL-Cf-4640_at	0	100	2.9	8.32E-03		
GL-Cf-5303_at	100	100	−2.45	3.10E-02	E:3ge,N:3ge	LOC485993
GL-Cf-5412_at	100	100	−2.09	1.53E-02	E:3ge	GPT2
GL-Cf-6613_at	100	100	−2.54	2.80E-02	E:g1,tr,N:g1,tr	LOC479911
GL-Cf-7513_at	100	100	−2.25	4.01E-02	E:3ge,N:3ge	LOC475819
GL-Cf-778_at	100	100	−2.68	2.05E-02	E:g1,N:g1,tr	LOC479818
GL-Cf-7969_at	100	33.33	−1.98	2.78E-02	E:g1,N:g1,tr	LOC485528

Table 3
(Continued)

Probeset	% present vehicle	% present phenobarbital	FC signed magnitude (vehicle vs. phenobarbital)	t-test p value (vehicle vs. phenobarbital)	Native locus match	Native locus symbol
GL-Cf-8031_at	100	100	−1.93	3.53E-02	E:g1,tr,N:g1,tr	LOC475476
GL-Cf-8096_s_at	100	100	2.32	1.44E-02	E:g1,tr,N:g1,tr	PLA2G7
GL-Cf-8208_at	100	100	2.23	8.77E-03	E:3ge,N:g1,tr	LOC610502
GL-Cf-827_at	100	100	2.12	4.52E-02	E:g1,tr,N:g1,tr	LOC475814
GL-Cf-8582_at	100	100	2.69	4.14E-06	E:3ge,N:3ge	LOC612549
GL-Cf-870_at	100	100	1.82	2.89E-02		
GL-Cf-9013_r_at	100	100	2.85	6.34E-03		
GL-Cf-9219_at	100	100	1.95	2.95E-02	E:g0,N:g1,tr	Cyp3a13 Cyp3a18 LOC489851
GL-Cf-9481_at	66.67	66.67	2.19	8.43E-03	E:g1,tr,N:g1,tr	LOC487820
GL-Cf-9493_at	100	100	6.7	2.17E-03	E:g1,N:tr	AKR1C3 PGFS
GL-Cf-958_i_at	100	100	2.74	4.97E-02	N:g1	CYP2D15 CYP2D

Affymetrix probesets are listed that met the statistical criteria noted in the text (mean group values ≥1.8-fold difference between control and phenobarbital groups with *t*-test *p* values <.05 and a 100% present call in either the control or phenobarbital group). Other information and definitions of abbreviations are provided in the Table 1 footnote.

Table 4

Probeset	Native locus symbol	Orthology	Reliability	Gene symbol	Associated organism	RAE230_2 probesets
GL-Cf-10423_at	Fads3	E:UBRH	Low	Fads3	Rat	1372476_at (High)
GL-Cf-1043_at	LOC489393	H:unique, E:UBRH	High	Apoa5	Rat	1369011_at (High)
GL-Cf-10437_at	LOC607852	H:unique, E:UBRH	High	Acly	Rat	1367854_at (High) 1395841_at (Low)
GL-Cf-1049_at	LOC479182	H:unique, E:UBRH	Medium	Hmgcr	Rat	1375852_at (High) 1387848_at (High)
GL-Cf-1062_at	PPP2CA	H:unique, E:UBRH	Medium	Ppp2ca	Rat	1379616_at (Low) 1388805_at (Low) 1398790_at (High)
GL-Cf-10667_at	SERPINE1 PAI-1	H:unique, E:UBRH	High	4833436O05	Mouse	1369493_at (High)
GL-Cf-11061_at	LOC479363	H:unique, E:UBRH	Low	Prlr	Rat	1370384_a_at (Medium) 1370789_a_at (Medium) 1392612_at (Low) 1376944_at (Low)
GL-Cf-11636_at	LOC610935	H:unique, E:UBRH	Medium	Nqo1	Rat	1387599_a_at (High)
GL-Cf-11716_at	LOC475225	H:unique, E:UBRH	High	Cyp51	Rat	1367979_s_at (High) 1387020_at (Medium) 1396961_at (Low)
GL-Cf-11757_r_at						
GL-Cf-12118_at						
GL-Cf-12207_at	LOC607852	H:unique	Medium	Acly	Rat	1367854_at (High) 1395841_at (Low)
GL-Cf-12295_at	LOC612549	H:unique, E:UBRH	Low	C1d	Mouse	1397387_at (Low) 1377731_at (High)

Table 4
(Continued)

Probeset	Native locus symbol	Orthology	Reliability	Gene symbol	Associated organism	RAE230_2 probesets
GL-Cf-12678_at	FBN1	H:unique, E:UBRH	Medium	Fbn1	Rat	1368829_at (Low) 1387351_at (High)
GL-Cf-13327_at						
GL-Cf-1344_at	LOC476600	H:unique, E:UBRH	High	Alas1	Rat	1367982_at (Medium)
GL-Cf-13448_at						
GL-Cf-13922_at	LOC475926 LOC607806	E:UBRH, H:unique	Ambiguous	Multiple(Cbr2;Dcxr)	Rat, mouse	
GL-Cf-13983_at	LOC612198	H:unique, E:UBRH	Low	KIAA1423	Human	1384428_at (Medium)
GL-Cf-14020_at	LOC487139	H:unique, E:UBRH	Medium	Pfkfb3	Rat	1369794_a_at (High) 1390391_at (Low) GL-Cf-14020_at 1397082_at (Medium)
GL-Cf-14153_at	IL1RN IL-1ra	H:unique, E:UBRH	Medium	Il1rn	Rat	1387835_at (High)
GL-Cf-14195_at	LOC474666	H:unique, E:UBRH	High	SLC27A6	Human	
GL-Cf-1575_at	LOC475939	H:unique, E:UBRH	Medium	Hsd17b12	Rat	1368051_at (High)
GL-Cf-1587_at	LOC478665	H:unique, E:UBRH	High	Fetub	Rat	1387082_at (High)
GL-Cf-1937_at	SEC22A	E:UBRH	Medium	Sec22l2	Rat	
GL-Cf-1940_at	CYP3A26	E:RHS, H:multiple	Ambiguous	Multiple(Cyp3a13; Cyp3a44; Cyp3a41;Cyp3a11)	Rat, mouse	
GL-Cf-1979_at	Fads3 LOC483793	E:UBRH, H:unique	Ambiguous	Multiple(Fads1;Fads3)	Rat	
GL-Cf-2178_at	LOC479147	H:unique, E:UBRH	High	Riok2	Rat	1377503_at (High) 1394878_at (Low)

Probe	Gene ID	Mapping	Conf.	Gene	Species	Affymetrix
GL-Cf-2228_at	LOC478664	H:unique, E:UBRH	High	Dnajb11	Rat	1389308_at (High)
GL-Cf-235_at	DVS27	H:unique, E:UBRH	High	RGD1311155	Rat	1373970_at (High)
GL-Cf-274_at	PLA2G7	H:unique, E:UBRH	High	Pla2g7	Rat	1389581_at (High)
GL-Cf-2756_at						
GL-Cf-3098_at	LOC488291	H:unique, E:UBRH	High	Pir	Rat	1377662_at (High)
GL-Cf-380_at	CXADR	H:unique, E:UBRH	High	Cxadr	Rat	1374273_at (Medium) 1384816_at (High)
GL-Cf-3826_at	LOC475175	H:unique, E:UBRH	High	Afm	Rat	1371266_at (High)
GL-Cf-3844_at	LOC474802	H:unique, E:UBRH	High	Ltb4dh	Rat	1388102_at (High)
GL-Cf-4182_at	LOC480220	H:unique, E:UBRH	High	Myo5b	Rat	1368355_at (High)
GL-Cf-4202_at	LOC483378	H:unique, E:UBRH	High	Fasn	Rat	1367707_at (Medium) 1367708_a_at (Medium)
GL-Cf-4324_at	LOC475225	H:unique, E:UBRH	High	Cyp51	Rat	1367979_s_at (High) 1387020_at (Medium) 1396961_at (Low)
GL-Cf-4640_at						
GL-Cf-5303_at	LOC485993	H:unique, E:UBRH	Low	Slc25a15	Rat	1374800_at (Low) 1381124_at (Low) 1393342_at (Low) 1393947_at (High) 1397647_at (High)
GL-Cf-5412_at	GPT2	E:UBRH	Low	Gpt1	Rat	1387052_at (High)
GL-Cf-6613_at	LOC479911	H:unique, E:UBRH	High	Gstm5	Rat	1370813_at (High)
GL-Cf-7513_at	LOC475819	H:unique, E:UBRH	Low	Fmo5	Rat	1383248_at (Medium)
GL-Cf-778_at	LOC479818	H:unique, E:UBRH	High	Crym	Rat	1368059_at (Medium)
GL-Cf-7969_at	LOC485528	H:unique, E:UBRH	Medium	Ranbp5	Mouse	
GL-Cf-8031_at	LOC475476	H:unique, E:UBRH	High	Tdo2	Rat	1368720_at (High)

Table 4
(Continued)

Probeset	Native locus symbol	Orthology	Reliability	Gene symbol	Associated organism	RAE230_2 probesets
GL-Cf-8096_s_at	PLA2G7	H:unique, E:UBRH	High	Pla2g7	Rat	1370319_at (High)
GL-Cf-8208_at	LOC610502	H:unique, E:UBRH	Medium	Ppif	Rat	
GL-Cf-827_at	LOC475814	H:unique, E:UBRH	High	Hao2	Rat	1387139_at (High)
GL-Cf-8582_at	LOC612549	H:unique, E:UBRH	Low	C1d	Mouse	
GL-Cf-870_at						
GL-Cf-9013_r_at						
GL-Cf-9219_at	Cyp3a13 Cyp3a18 LOC489851	H:unique, E:UBRH	Ambiguous	Multiple (Cyp3a13; Cyp3a18)	Rat	
GL-Cf-9481_at	LOC487820	H:unique, E:UBRH	High	Paqr3	Rat	
GL-Cf-9493_at	AKR1C3 PGFS	E:UBRH, H:unique	Ambiguous	Multiple (LOC171516; Akr1c6)	Rat	
GL-Cf-958_i_at	CYP2D15 CYP2D	H:unique	Medium	Cyp2d22	Rat	1387913_at (High) 1370329_at (Medium)

Affymetrix probesets are listed that met the statistical criteria noted in the text (mean group values ≥1.8-fold difference between control and phenobarbital groups with *t*-test *p* values <.05 and a 100% present call in either the control or phenobarbital group). Other information and definitions of abbreviations are provided in the **Table 2** footnote.

Canine Ensembl Gene IDs were similarly translated via Ensembl's orthology tables to rat gene IDs when possible, otherwise to mouse or human. The Ensembl orthology method (Unique Best Reciprocal Hit (UBRH), Multiple Best Reciprocal Hit (MBRH), Reciprocal Hit based on Synteny (RHS)) was recorded as a quality metric. These Ensembl Gene IDs were then translated to Entrez Gene IDs via the correspondences available in Ensembl (when possible) and, finally, the results of these two paths, Ensembl and NCBI-based, were compared. When every step of both paths was of the highest possible quality and the two paths converged on a unique answer, the result was recorded as high reliability. This requires a genome match to a known exon and a match to a transcript from that gene, a unique orthology call from HomoloGene and a UBRH orthology call from Ensembl. Where an association relied on a low-reliability method and was not confirmed by another method, the association was recorded as low reliability. High-reliability associations via only a single path were recorded as medium reliability. Where more than one association was possible, they are both recorded as ambiguous.

A similar process was used to map rat RAE230_2 probesets into rat (or when necessary mouse or human loci) and where these loci occur in **Table 1** to **Table 4**, the rat probeset names are given in the final column along with the reliability of their association with the locus.

3.2. Probeset Level Cross-Genomic Annotation

As a supplemental method, we also took the path illustrated by dotted arrows in **Fig. 1**. We used BLAST (ver. 2.2.13) to compare each target sequence with all transcripts in the RefSeq database. Matches were declared for an High Scoring Pair (HSP) score >22 and an e-value <0.005. Each of the matching RefSeq transcript sequences was then associated with the corresponding Entrez Gene ID (as above) and looked up in HomoloGene where a matching rat Entrez Gene ID was recorded (or, in the absence of rat, mouse or human). Again, the uniqueness of the match was recorded as a quality metric. Where these matches gave identical results to the above, they were suppressed. Where they supplemented a low-reliability result with additional support, or identified a new association, they are recorded in the "Non-native match" column of **Table 2**. So, for example, probeset GL-Cf-6542_at, which the native locus method found to be near but not overlapping with the same gene (the "3ge" method), was confirmed by a match to a human transcript from the orthologous locus and therefore allowed the reliability score to be raised from "Low" to "Medium." Similarly, GL-Cf-261_at could not be associated with a known canine locus at all by any of the methods used above, but could be associated with a human transcript and then by orthology into rat. These cases, and other similar ones recorded in **Table 2** illustrate the point that direct comparisons to transcripts or annotated genomes from more mature genome sequences can yield significant dividends in increasing the annotation coverage. They do this, however, at the cost of less rigorous association criteria, which renders their reliability at least somewhat lower.

4. Experience with Cross-Species Microarray Experiments

One of the best-studied responses in toxicology is that of the rodent liver to agents such as phenobarbital *(16)*. In rats, treatment with this compound increases the expression in liver of a number of genes coding for enzymes that carry out metabolic conversion of endogenous and exogenous chemicals. Most notably, the transcript levels for cytochrome P450 2B (Cyp2B) are increased several hundred fold *(17)*. Phenobarbital induction of cytochrome P450s has been described in several species, and in dog this results in the induction of both Cyp3A and Cyp2B activities and immunoreactive isoforms *(18,19)*, the latter being identified as CYP2B11 *(20)*.

We studied the effect of phenobarbital treatment on global transcript levels in canine liver using a custom Affymetrix GeneChip microarray. Male beagles (three animals/group) were treated with either vehicle (water) or phenobarbital, the latter dosing regime being 10 mg kg^{-1} day^{-1} for 2 days and 30 mg kg^{-1} day^{-1} for 5 days. Significantly regulated genes were identified as those with magnitude signed fold-change between mean group values ≥ 1.8 with p value $<.05$, and of the 12,506 probesets, 57 met this criteria. Using the "genome to genome through orthology" mapping method described above, 44 of the 57 "dysregulated probesets" were unambiguously annotated, representing 40 unique transcripts (**Table 2**). Not surprisingly, this list included several putative canine cytochrome P450 transcripts, notably Cyp51, Cyp3A and Cyp2D, as well as the known canine transcript CYP3A26. The known canine transcript CYP2B11 is indeed represented on the microarray but is upregulated only 1.37 fold (**Table 1**), below the fold-change filter. This is in contrast with not only results for induction of Cyp2B mRNA in rats but also with the observation by Graham et al. that CYP2B activity is induced many fold in the livers of phenobarbital-treated dogs *(18)*. The unfiltered results for a number of probesets representing other cytochrome P450 transcripts are given in **Table 1**.

The modest results here beg the question of how the response of the dog to phenobarbital compares with that of the rat. If one were to comprehensively compare the transcripts regulated by phenobarbital in this study in dogs with those regulated by phenobarbital in rats, one would first have to consider what transcripts are monitored by the arrays used for the two species. With a small result set such as described above, the task is simple and can be carried out with various manual approaches. However, if the intent is to compare and contrast gene regulation on a global basis (overall pathway level with a large gene list) between species, the global level of transcript coverage overlap must be determined and here an automated method is required, with such methods being described in the previous section.

On the other hand, if the results for the same transcript are discordant in two species, this may be due simply to the poor performance of the corresponding probe on one microarray. In the case noted above (phenobarbital induction of CYP2B in dogs), the signal for the probeset was significant in both control and treated liver samples (data not shown). Were that not the case, an alternative mRNA measurement (e.g., qRT-PCR) may be considered to confirm a difference in transcriptional response. Automated methods could be envisioned that would compare probe signal levels for homologous transcripts on different microarrays, but such methods would not necessarily be suitable for comparing probe performance for low-abundance transcripts.

The preceding discussion highlights the third issue, namely whether discordant results for the same transcript in two species are truly due to differences in the response of one species or whether the particular experiment was not representative of the true response. Obviously, this question can be addressed with further experiments replicating and/or altering the original conditions.

Finally, even with an observation of similar transcriptional responses in two species for a particular homologous gene, the question remains, as noted in previous sections, as to whether the transcriptional response is translated into protein and functional activity. As noted above, Graham et al. had observed in their study an increase in both immunoreactive protein and activity *(18)*. On the other hand, Duignan and colleagues *(21)* have already observed that the rat and canine CYP2B proteins differ somewhat in catalytic specificities.

5. Conclusion

This discussion highlights some of the limitations of current efforts in comparing global transcriptional responses across species. A recurring problem, not just for cross-species analysis but for all microarray analysis, is that of incomplete functional annotation for the genes even in one species. Resources such as Entrez Gene *(22)* provide a great deal of such annotation, particularly with Reference Into Function (GeneRIF), which incorporates annotation linked to published reports, but such information is provided only through the initiative of individual contributors. Furthermore, this information is not compiled into a keyword format, nor does it incorporate information that may be included in a protein record. And whereas the HomoloGene resource does provide information about gene homologs in several species, it also lacks functional information that may be found in Entrez Gene or in protein records. Ensembl provides alternatives that also have significant strengths, but mapping between the Entrez Gene and Ensembl worlds can be highly problematic.

One hope is that pathway analysis may provide a way to focus comparisons across species, yet such analyses still have limitations *(23)*. As noted above, a critical problem is the further annotation that will link genes from relevant species into pathways. Of course, not only does this require more research, but also it is complicated by the fact that many proteins can have extremely diverse functions; for example, glyceraldehyde-3-phosphate dehydrogenase, a presumed "housekeeping gene," has been found to have roles in apoptosis and neurodegenerative diseases *(24)*.

Even with these challenges, a comparison of transcriptional responses across species still offers huge impact in the drug discovery and development process, not only in identifying appropriate disease or pharmacology models but also in extrapolating the human risk of responses seen in nonclinical safety studies.

References

1. Birch, H. E., and Schreiber, G. (1986) Transcriptional regulation of plasma protein synthesis during inflammation. *J. Biol. Chem.* **261**, 8077–8080.
2. Wegenka, U. M., Buschmann, J., Lutticken, C., Heinrich, P. C., and Horn, F. (1993) Acute-phase response factor, a nuclear factor binding to acute-phase response elements, is rapidly activated by interleukin-6 at the posttranslational level. *Mol. Cell. Biol.* **13**, 276–288.
3. Fitch, W. M. (1970) Distinguishing homologous from analogous proteins. *Syst. Zool.* **19**, 99–113.
4. Chen, F., Mackey, A. J., Stoeckert, C. J., Jr., and Roos, D. S. (2006) OrthoMCL-DB: querying a comprehensive multi-species collection of ortholog groups. *Nucleic Acids Res.* **34**, D363–368.
5. Li, L., Stoeckert, C. J., Jr., and Roos, D. S. (2003) OrthoMCL: identification of ortholog groups for eukaryotic genomes. *Genome Res.* **13**, 2178–2189.
6. Remm, M., Storm, C. E., and Sonnhammer, E. L. (2001) Automatic clustering of orthologs and in-paralogs from pairwise species comparisons. *J. Mol. Biol.* **314**, 1041–1052.
7. Doehmer, J., Holtkamp, D., Soballa, V., Raab, G., Schmalix, W., Seidel, A., et al. (1995) Cytochrome P450 mediated reactions studied in genetically engineered V79 Chinese hamster cells. *Pharmacogenetics* **5** (Spec. No.), S91–96.
8. Glatt, H., Engelke, C. E., Pabel, U., Teubner, W., Jones, A. L., Coughtrie, M. W., et al. (2000) Sulfotransferases: genetics and role in toxicology. *Toxicol. Lett.* **112–113**, 341–348.
9. Wheeler, D. L., Barrett, T., Benson, D. A., Bryant, S. H., Canese, K., Chetvernin, V., et al. (2006) Database resources of the National Center for Biotechnology Information. *Nucleic Acids Res.* **34**, D173–180.

10. Birney, E., Andrews, T. D., Bevan, P., Caccamo, M., Chen, Y., Clarke, L., et al. (2004) An overview of Ensembl. *Genome Res.* **14**, 925–928.
11. Blake, J. A., Eppig, J. T., Bult, C. J., Kadin, J. A., and Richardson, J. E. (2006) The Mouse Genome Database (MGD): updates and enhancements. *Nucleic Acids Res.* **34**, D562–567.
12. O'Brien, K. P., Remm, M., and Sonnhammer, E. L. (2005) Inparanoid: a comprehensive database of eukaryotic orthologs. *Nucleic Acids Res.* **33**, D476–480.
13. Johnson, R. J., Williams, J. M., Schreiber, B. M., Elfe, C. D., Lennon-Hopkins, K. L., Skrzypek, M. S., and White, R. D. (2005) Analysis of gene ontology features in microarray data using the Proteome BioKnowledge Library. *In Silico Biol.* **5**, 389–399.
14. Hodges, P. E., Carrico, P. M., Hogan, J. D., O'Neill, K. E., Owen, J. J., Mangan, M., et al. (2002) Annotating the human proteome: the Human Proteome Survey Database (HumanPSD) and an in-depth target database for G protein-coupled receptors (GPCR-PD) from Incyte Genomics. *Nucleic Acids Res.* **30**, 137–141.
15. Kent, W. J. (2002) BLAT–the BLAST-like alignment tool. *Genome Res.* **12**, 656–664.
16. Waxman, D. J., and Azaroff, L. (1992) Phenobarbital induction of cytochrome P-450 gene expression. *Biochem J.* **281** (Pt 3), 577–592.
17. Omiecinski, C. J., Walz, F. G., Jr., and Vlasuk, G. P. (1985) Phenobarbital induction of rat liver cytochromes P-450b and P-450e. Quantitation of specific RNAs by hybridization to synthetic oligodeoxyribonucleotide probes. *J. Biol. Chem.* **260**, 3247–3250.
18. Graham, R. A., Downey, A., Mudra, D., Krueger, L., Carroll, K., Chengelis, C., et al. (2002) In vivo and in vitro induction of cytochrome P450 enzymes in beagle dogs. *Drug Metab. Dispos.* **30**, 1206–1213.
19. Jayyosi, Z., Muc, M., Erick, J., Thomas, P. E., and Kelley, M. (1996) Catalytic and immunochemical characterization of cytochrome P450 isozyme induction in dog liver. *Fundam. Appl. Toxicol.* **31**, 95–102.
20. Graves, P. E., Elhag, G. A., Ciaccio, P. J., Bourque, D. P., and Halpert, J. R. (1990) cDNA and deduced amino acid sequences of a dog hepatic cytochrome P450IIB responsible for the metabolism of 2,2′,4,4′,5,5′-hexachlorobiphenyl. *Arch Biochem. Biophys.* **281**, 106–115.
21. Duignan, D. B., Sipes, I. G., Ciaccio, P. J., and Halpert, J. R. (1988) The metabolism of xenobiotics and endogenous compounds by the constitutive dog liver cytochrome P450 PBD-2. *Arch Biochem. Biophys.* **267**, 294–304.
22. Maglott, D., Ostell, J., Pruitt, K. D., and Tatusova, T. (2005) Entrez Gene: gene-centered information at NCBI. *Nucleic Acids Res.* **33**, D54–58.
23. Khatri, P., and Draghici, S. (2005) Ontological analysis of gene expression data: current tools, limitations, and open problems. *Bioinformatics* **21**, 3587–3595.
24. Chuang, D. M., Hough, C., and Senatorov, V. V. (2005) Glyceraldehyde-3-phosphate dehydrogenase, apoptosis, and neurodegenerative diseases. *Annu. Rev. Pharmacol. Toxicol.* **45**, 269–290.

9

Toxicogenomics in Biomarker Discovery

Marc F. DeCristofaro and Kellye K. Daniels

Summary

In the area of toxicology, the subdiscipline of toxicogenomics has emerged, which is the use of genome-scale mRNA expression profiling to monitor responses to adverse xenobiotic exposure. Toxicogenomics is being investigated for use in the triage of compounds through predicting potential toxicity, defining mechanisms of toxicity, and identifying potential biomarkers of toxicity. Whereas various approaches have been reported for the development of algorithms predictive of toxicity and for the interpretation of gene expression data for deriving mechanisms of toxicity, there are no clearly defined methods for the discovery of biomarkers using gene expression technologies. Ways in which toxicogenomics may be used for biomarker discovery include analysis of large databases of gene expression profiles followed by *in silico* mining of the database for differentially expressed genes; the analysis of gene expression data from preclinical studies to find differentially expressed genes that correlate with pathology (coincident biomarker) or precede pathology (leading biomarker) within a lead series; or gene expression profiling can be performed directly on the blood from preclinical studies or clinical trials to find biomarkers that can be obtained noninvasively. This chapter broadly discusses the issues and the utility of applying toxicogenomics to biomarker discovery.

Key Words: biomarker; blood; gene expression; heart; kidney; molecular profiling; toxicogenomics; transcriptomics.

1. Introduction

After the release of the rough draft of the human genome in early 2001 *(1,2)*, the fields of genetics and genomics progressed rapidly, with new tools and technologies being used to identify and to quantify global gene expression changes. The advent of these technologies has advanced the understanding

From: *Methods in Molecular Biology, vol. 460: Essential Concepts in Toxicogenomics*
Edited by: D. L. Mendrick and W. B. Mattes © Humana Press, Totowa, NJ

of the distribution, function, and regulation of genes from the single cell to the systems biology level and is expanding the ways in which biological research asks and answers questions. In the area of toxicology, the subdiscipline of toxicogenomics has emerged, which is the use of genome-scale mRNA expression profiling (i.e., transcriptomics) to monitor responses to adverse xenobiotic exposure. Toxicity profiling with DNA microarrays has led to the discovery of better descriptors of toxicity, toxicant classification, and exposure monitoring than current indicators. One major goal in transcript profiling is the development of biomarkers and signatures of chemical toxicity. The introduction of toxicogenomics databases supports researchers in sharing, analyzing, visualizing, and mining expression data, assists the integration of transcriptomics, proteomics, and toxicology data sets, and may permit *in silico* biomarker and signature pattern discovery.

Molecular profiling methods are used to monitor global cellular responses due to drug treatment or toxicity perturbations. These experiments yield unique molecular fingerprints, which reflect the cumulative response of complex interactions. If these interactions can be significantly correlated with an end point, the molecular fingerprint may be termed a *biomarker*, meaning a characteristic that is objectively measured and evaluated as an indicator of normal biological or pathogenic processes or a pharmacological response to a therapeutic intervention *(3)*. Currently, there are no defined methods for biomarker discovery using toxicogenomics; even though there is a need for sensitive, specific, predictive, and reproducible biomarkers *(4)* across a multitude of drug-induced disease-based organ toxicities. Therefore, this chapter will focus on describing the application of the principles of toxicogenomics to biomarker discovery.

2. Materials

Toxicogenomics is being investigated for use in the triage of compounds by attempting to predict potential toxicity, define mechanisms of toxicity, and identify potential biomarkers of toxicity. Currently, there are no clearly defined methods for the discovery of biomarkers using gene expression technologies. The basic assumption is that most toxicologic end points will be associated with changes in gene expression *(5)* and that these changes will precede observable histopathology. Because of this and the sensitivity of genome-wide gene expression technologies, it is believed that they have the potential to provide information at earlier times and at lower doses than traditional toxicologic end points. Most treatment effects are not accounted for by a single biomarker because of the complexity of disease and compound-related toxicity. The use of patterns of multiple biomarkers provided by genomics-based approaches may prove to be more accurate than traditional approaches, because they could represent different components or pathways that lead to the compound-induced toxicity *(3,6)*. The strength of gene expression profiling is that it is system wide and does not rely on an individual gene but on a pattern of genes that are different in the toxic situation when compared with control groups *(7)*. The ideal genomic marker would be one where the intensity of dysregulation compared with control increases with the damage *(8)*.

F.M. Goodsaid suggested that a predictive biomarker have (1) expression before observable pathology, (2) correlation between inception and development of pathology, (3) dose-responsiveness, and (4) correlation with removal of the toxicant *(8)*. Gene expression data integrated with classic toxicology end points should provide a more informed and comprehensive decision *(5)*.

In order for a biomarker to be useful, it must be sensitive, specific, predictive, and reproducible *(4,7)*. One way to approach biomarker discovery is the analysis of large databases of gene expression profiles populated with compounds resulting in defined, well-known adverse effects followed by *in silico* mining of the database for differentially expressed genes *(4)*. Ideally, these genes would correlate with histopathologic analysis and could be considered more general markers of toxicity as they would not be compound specific. This method becomes more effective as the number of compounds and different types of histopathology are profiled. The biggest downfall to this approach is tied to the cost associated with the development of the reference database.

Another approach is the analysis of gene expression data from preclinical studies to find differentially expressed genes that correlate with pathology (coincident biomarker) or precede pathology (leading biomarker) within a lead series. These markers could be more specific to a development project. Biomarkers for use in toxicity are only helpful if they are able to accurately predict toxicity before traditional toxicology methods. Because most traditional toxicology methods are designed to result in observable toxicity after 14 days, 1- and 3-day time points are being investigated for genomics biomarker discovery. With this shortened time frame, the gene expression changes would be primarily compound-related and not the result of complicating secondary events, and a potentially huge cost savings in time and in the materials needed for 14 days of animal testing would result *(4)*. One complicating factor is the lack of annotation for some of the most common preclinical animal model species, specifically the rat and the canine. Genomics-based biomarker identification is particularly difficult in the canine, where the number of annotated sequences is very limited *(4,8)*. This should change in the future.

Biomarkers discovered in preclinical studies can be found by assaying the target tissue and strict validation is not essential. For biomarkers intended for clinical studies, the tissue must be obtained noninvasively, and strict validation is needed. This does not mean that there is no value in the preclinical studies.

Biomarkers discovered in the target tissue from preclinical studies may provide mechanistic information that could aid in the discovery of markers in biological fluids. Alternatively, potential biomarkers discovered in preclinical studies could be triaged based on their accessibility in biological fluids and their correlation between mRNA and protein levels *(3,4,6)*. This assumes that once a potential genomics-based biomarker is discovered, the clinical assay developed based on this information would be protein based. Screening of various protein signatures from biological fluids in the clinic has been suggested as a goal in the future *(5)*.

Alternatively, gene expression profiling can be performed directly on the blood from preclinical studies or clinical trials to find biomarkers that can be obtained noninvasively. This would eliminate the need to triage potential biomarkers based on their accessibility. Genomics-based clinical studies may also be able to identify biomarkers that differentiate populations within a clinical trial. In theory, biomarkers can be used to separate responders from nonresponders *(4)*. It has been suggested that gene expression profiling of clinical samples could lead to novel indications and/or mechanisms *(5)*.

Validation of genomics-based markers should be done with alternative technologies. Because mRNA levels do not always correlate with protein levels, genomic biomarkers intended for clinical development should ultimately be validated using techniques such as quantitative RT-PCR, immunohistochemistry (IHC), and/or *in situ* hybridization (ISH). These technologies are more specific than genome-wide genomics-based technologies and can aid in the corroboration of the mRNA data *(5)*.

3. Methods

3.1. Kidney

Renal injury and clinical renal failure can occur through a variety of mechanisms after administration of different therapeutic agents. The antiviral agent acyclovir has been described to induce nephropathy by crystal deposition *(9)*. The epithelial cells of the proximal tubules of the kidney cortex are the primary target for chloroform-induced cytotoxicity and regenerative cell proliferation in the kidney *(10)*. Mercuric chloride ($HgCl_2$) induces dose-dependent alterations in the cytoplasm and the nucleus of proximal tubule cells resulting in acute renal failure *(11,12)*. Puromycin aminonucleoside (PAN) causes proteinuria by affecting the glomerular epithelial cell surface layer *(13)*, and indomethacin causes phospholipid accumulation in the papilla of the rat kidney *(14)*.

Current kidney functional tests are either insensitive, variable, or nonspecific to kidney injury *(15,16)*. Hewitt et al. *(16)* have expressed the need for markers of renal dysfunction that could be used to detect damage earlier than is currently available or to follow the progression of toxicity. Biomarkers discovered using gene expression profiles may be able to aid in these areas. The literature has shown that the exposure to a nephrotoxicant results in the dysregulation of many genes. These markers of kidney injury include clusterin *(17)*, fibrinogen alpha, beta, and gamma polypeptides *(18)*, kidney injury molecule 1 (KIM-1), lipocalin 2, secreted phosphoprotein 1 (osteopontin) *(15,19)*, and tissue inhibitor of metalloproteinase 1 (TIMP1) *(20)*. RT-PCR and ISH have demonstrated a downregulation of calbindin at the mRNA and protein level in rat kidneys after cyclosporin A treatment *(5,21)*. However, the dysregulation seen with most, if not all, of these markers tends to be concurrent with observable histopathology. In order for genomics-based biomarkers to add value, they should be markers that are dysregulated prior to the observed histopathology. Future experiments should be designed using earlier time points specifically for the discovery of biomarkers that accurately predict toxicity before traditional toxicology methods at 14 days. Only then will genomics-based biomarker discovery be able to fulfill its promise.

3.2. Heart

There are examples in almost every therapeutic class of drugs causing unanticipated cardiotoxicity, including the anthracyclines *(22)*, trastazumab (Herceptin, Genentech, South San Francisco, CA) *(23)*, and other chemotherapeutic agents *(24)*. Recent attention has focused on the COX-2 selective nonsteroidal anti-inflammatory drugs (NSAIDs) such as valdecoxib (Bextra, Pfizer, New York, NY), for which the Food and Drug Administration requested voluntary withdrawal because of the potential for serious cardiovascular adverse events *(25)*. Likewise, a typical effect of class III antiarrhythmic drugs, like amiodarone, is the prolongation of the electrocardiogram QT interval through blockade of the human ether-à-go-go-related gene (hERG) potassium channel, increasing the likelihood of development of a polymorphous ventricular arrhythmia *(26)*. Robust biomarkers, which can reliably predict adverse drug effects, especially those that can bridge preclinical and clinical settings, are particularly valuable for ensuring the safety of new drug molecules.

According to Wallace and colleagues *(27)*, drug-induced cardiac injury may be categorized according to *(1)* structural damage, *(2)* functional deficits that may or may not be associated with histopathologic changes, or *(3)* altered cell or tissue homeostasis in the absence of obvious structural or functional deficits. Myocardial infarction represents the prototypical phenotype for which biomarkers have been developed and used in diagnosis, risk stratification, and treatment for several decades. The process of myocardial necrosis in myocardial infarction leads to the release of proteins from dead myocytes into circulation, providing relatively accessible targets for biomarker(s) identification. Numerous serum biomarkers are either in use or have been proposed for use in the diagnosis of myocardial injury, including acute myocardial infarction. These include the myocardial isoenzyme of creatinine kinase (CK-MB), lactate dehydrogenase (LDH), myoglobin, heart fatty acid–binding protein (FABP), and troponins T (cTnT) and I (cTnI) *(27,28)*. Elevated serum cTnT and cTnI levels are widely implemented biomarkers of cardiovascular injury; however, they are not biomarkers for all forms of drug-induced cardiac injury. Cardiac cell injury that does not result in altered cardiac muscle cell membrane permeability may not be associated with increases in serum troponins, and serum troponins are not expected to increase in direct response to drug-induced arrhythmias, altered ion homeostasis, valvular disease, contractile dysfunction, and so forth *(27)*. Because of this, advances in molecular genetics and biology have shifted the paradigm for identification of markers from large-scale epidemiologic studies to studies on genomic- and proteomic-based techniques. Although cardiac myocytes are considered terminally differentiated cells, they can and do exhibit significant gene expression alterations in response to stress *(29)*. During pathologic cardiac hypertrophy, for example, fetal genes not normally expressed in the adult heart are re-expressed, including β-myosin heavy chain and α-skeletal actin *(30,31)*, along with induction of immediate-early genes like c-*fos*, c-*jun*, and c-*myc* *(32,33)*. The gene dysregulation seen with most, if not all, of these markers tends to be concurrent with observable histopathology. For a gene-based marker to be of utility, first, its protein product needs to be secreted or released from the cell after injury. Second, the time course of expression alteration after drug treatment must precede changes in clinical chemistry and histopathology. Third, normal distribution of expression across multiple species, including rat, canine, and human tissues, should be restricted to the tissue of interest. Fourth, the expression of the proposed biomarker should be restricted to the disease state for which it is targeted.

The application of toxicogenomics to cardiovascular biomarker discovery offers the opportunity to provide insight into the mechanisms of disease manifestation and progression, as well as the identification of a number of candidate biomarkers for all forms of cardiovascular diseases. The field is in the early stages of evolution, and large-scale clinical studies are required to validate the utility of newly identified biomarkers in diagnosis, risk stratification, and treatment of cardiovascular diseases.

3.3. Liver

Although every organ system can be affected by drug-induced toxicity, the most frequent site is the liver. Hepatotoxicity is the most common reason for the withdrawal of an approved drug from the market, and more than 600 drugs have been associated with hepatotoxicity. It is currently the reason for more than 50% of cases of acute liver failure in the United States *(34,35)*.

The most widely used clinical chemistry tests for liver injury are serum alanine aminotransferase (ALT) and aspartate aminotransferase (AST) levels. These enzymes are present in hepatocytes and leak into the serum after hepatocyte damage. Although ALT is more sensitive than AST, neither is an ideal marker, as they do not always correlate with the extent of hepatic injury *(36,37)*. There is clearly a need for better biomarkers of hepatotoxicity.

Toxicogenomic-based biomarker discovery should lead to the discovery of more sensitive biomarkers. Genes such as heme oxygenase 1 *(38)*, lipopolysaccharide binding protein *(39)*, and lipocalin 2 *(40)* have been identified as markers of hepatocellular damage. Future genomics experiments designed specifically for the discovery of biomarkers should be able to provide better biomarkers of hepatotoxicity than are currently available.

3.4. Blood

Molecular signatures can be useful to guide basic research when obtained experimentally from animal tissues that are inaccessible in the clinic. Signatures can sometimes be obtained from patients by biopsy of target organs or tumors, but these will not be available from normal subjects in early-stage clinical trials. Signatures from blood are most accessible and would have the widest clinical usefulness *(41)*. Gene expression profiling from blood remains a challenge because blood is a complicated biological system consisting of a variety of cell types at different stages of development. Blood is also one of the most variable tissue types for gene expression analysis. Hence, the success of a blood microarray study depends on the choice of cell isolation method and preparation technique *(42)*. Methods for assessing gene expression from whole-blood samples that do not block globin are suboptimal, given the interference with accurate biological recognition of transcription events caused by high levels of globin. Affymetrix (Santa Clara, CA) has developed the Globin Reduction Protocol *(43)*, and Gene Logic (Gaithersburg, MD) has developed a proprietary methodology termed the Globin Interference Reduction Protocol (www.genelogic.com). The latter technique is effective on whole-blood total RNA regardless of isolation methods, works for isolation of RNA from different methods of blood preservation (e.g., Trizol or PAXgene), and is not limited to Affymetrix GeneChip arrays.

4. Conclusion

Toxicogenomics as a discipline is still evolving. Although there are no clearly defined methods for the discovery of biomarkers using gene expression technologies, the potential of this new field seems promising. The discovery of markers that precede pathology (leading biomarker) should be able to lessen the time and cost it currently takes to bring a new therapeutic to market. Future experiments, focusing on early time points and multiple doses, specifically designed for the discovery of biomarkers that accurately predict toxicity before traditional toxicology methods, will enable this discipline to fulfill its promise.

References

1. International Human Genome Sequencing Consortium. (2001) *Nature* **409**, 860–921.
2. Venter, J.C., Adams, M.D., Myers, E.W., Li, P.W., Mural, R.J., Sutton, G.G., et al. (2001) *Science* **291**, 1304–1351.
3. Biomarkers Definitions Working Group. (2001) Biomarkers and surrogate endpoints: preferred definitions and conceptual framework. *Clin. Pharmacol. Ther.* **69**, 89–95.
4. Bailey, W.J. and Ulrich, R. (2004) Molecular profiling approaches for identifying novel biomarkers. *Expert Opin. Drug Saf.* **3**, 137–151.
5. Guerreiro, N., Staedtler, F., Grenet, O., Kehren, J., and Chibout, S.D. (2003) Toxicogenomics in drug development. *Toxicol. Pathol.* **31**, 471–479.
6. Feng, Z., Prentice, R., and Srivastava, S. (2004) Research issues and strategies for genomic and proteomic biomarker discovery and validation: a statistical perspective. *Pharmacogenomics* **5**, 709–719.
7. Negm, R.S., Verma, M., and Srivastava, S. (2002) The promise of biomarkers in cancer screening and detection. *Trends Mol. Med.* **8**, 288–293.
8. Goodsaid, F.M. (2004) Identification and measurement of genomic biomarkers of nephrotoxicity. *J. Pharmacol. Toxicol. Methods* **49**, 183–186.
9. Perazella, M.A. (2003) Drug-induced renal failure: update on new medications and unique mechanisms of nephrotoxicity. *Am. J. Med. Sci.* **325**, 349–362.
10. Templin, M.V., Jamison, K.C., Wolf, D.C., Morgan, K.T., and Butterworth, B.E. (1996) Comparison of chloroform-induced toxicity in the kidneys, liver, and nasal passages of male Osborne-Mendel and F-344 rats. *Cancer Lett.* **104**, 71–78.
11. Stacchiotti, A., Borsani, E., Rodella, L., Rezzani, R., Bianchi, R., and Lavazza, A. (2003) Dose-dependent mercuric chloride tubular injury in rat kidney. *Ultrastruct. Pathol.* **27**, 253–259.
12. Stacchiotti, A., Lavazza, A., Rezzani, R., Borsani, E., Rodella, L., and Bianchi, R. (2004) Mercuric chloride-induced alterations in stress protein distribution in rat kidney. *Histol. Histopathol.* **19**, 1209–1218.
13. Kramer, J.A., Pettit, S.D., Amin, R.P., Bertram, T.A., Car, B., Cunningham, M., et al. (2004) Overview on the application of transcription profiling using selected nephrotoxicants for toxicology assessment. *Environ. Health Perspect.* **112**, 460–464.

14. Fernandez-Tome, M.C., and Sterin-Speziale, N.B. (1994) Short- and long-term treatment with indomethacin causes renal phospholipid alteration: a possible explanation for indomethacin nephrotoxicity. *Pharmacology* **48,** 341–348.

15. Han, W.K., Bailly, V., Abichandani, R., Thadhani, R., and Bonventre, J.V. (2002) Kidney Injury Molecule-1 (KIM-1): a novel biomarker for human renal proximal tubule injury. *Kidney Int.* **62,** 237–244.

16. Hewitt, S.M., Dear, J., and Star, R.A. (2004) Discovery of protein biomarkers for renal diseases. *J. Am. Soc. Nephrol.* **15,** 1677–1689.

17. Rosenberg, M.E. and Silkensen, J. (1995) Clusterin: physiologic and pathophysiologic considerations. *Int. J. Biochem. Cell. Biol.* **27,** 633–645.

18. Drew, A.F., Tucker, H.L., Liu, H., Witte, D.P., Degen, J.L., and Tipping, P.G. (2001) Crescentic glomerulonephritis is diminished in fibrinogen-deficient mice. *Am. J. Physiol. Renal Physiol.* **281,** F1157–1163.

19. Amin, R.P., Vickers, A.E., Sistare, F., Thompson, K.L., Roman, R.J., Lawton, M., et al. (2004) Identification of putative gene based markers of renal toxicity. *Environ. Health Perspect.* **112,** 465–479.

20. Chevalier, R.L. (2004) Biomarkers of congenital obstructive nephropathy: past, present and future. *J. Urol.* **172,** 852–857.

21. Grenet, O., Varela, M.C., Staedtler, F., and Steiner, S. (1998) The cyclosporine A-induced decrease in rat renal calbindin-D28kDa protein as a consequence of a decrease in its mRNA. *Biochem. Pharmacol.* **55,** 1131–1133.

22. Krischer, J.P., Epstein, S., Cuthbertson, D.D., Goorin, A.M., Epstein, M.L., and Lipshultz, S.E. (1997) Clinical cardiotoxicity following anthracycline treatment for childhood cancer. the pediatric oncology group experience. *J. Clin. Oncol.* **15,** 1544–1552.

23. Keefe, D.L. (2002) Trastuzumab-associated cardiotoxicity. *Cancer* **95,** 1592–1600.

24. Kruit, W.H., Punt, K.J., Goey, S.H., de Mulder, P.H., van Hoogenhuyze, D.C., Henzen-Logmans, S.C., and Stoter, G. (1994) Cardiotoxicity as a dose-limiting factor in a schedule of high dose bolus therapy with interleukin-2 and alpha-interferon. An unexpectedly frequent complication. *Cancer* **74,** 2850–2856.

25. Alert for Healthcare Professionals Valdecoxib (marketed as Bextra). http://www.fda.gov/cder/drug/InfoSheets/HCP/valdecoxibHCP.htm.

26. Recanatini, M., Poluzzi, E., Masetti, M., Cavalli, A., and De Ponti, F. (2005) QT prolongation through hERG K(+) channel blockade: current knowledge and strategies for the early prediction during drug development. *Med. Res. Rev.* **25,** 133–166.

27. Wallace, K.B., Hausner, E., Herman, E., Holt, G.D., MacGregor, J.T., Metz, A.L., et al. (2004) Serum troponins as biomarkers of drug-induced cardiac toxicity. *Toxicol. Pathol.* **32,** 106–121.

28. Marian, A.J. and Nambi, V. (2004) Biomarkers of cardiac disease. *Expert Rev. Mol. Diagn.* **4,** 805–820.

29. Kang, Y.J. (2001) Molecular and cellular mechanisms of cardiotoxicity. *Environ. Health Perspect.* **109** (Suppl 1)**,** 27–34.

30. Schwartz, L., Boheler, K.R., De La Bastie, D., Lompre, A.M., and Mercadier, J-J. (1992) Switches in cardiac muscle gene expresion as a result of pressure and volume overload. *Am. J. Physiol.* **262,** R364–369.
31. Rozich, J.D., Barnes, M.A., Schmid, P.G., Zile, M.R., McDermott, P.J., and Cooper, G. (1995) Load effects on gene expression during cardiac hypertrophy. *J. Mol. Cell. Cardiol.* **27,** 485–499.
32. Chien, K.R., Knowlton, K.U., Zhu, H., and Chien, S. (1991) Regulation of cardiac gene expression during myocardial growth and hypertrophy: molecular studies of an adaptive physiologic response. *FASEB J.* **5,** 3037–3046.
33. Kolbeck-Ruhmkorff, C. and Zimmer, H.G. (1995) Proto-oncogene expression in the isolated working rat heart: combination of pressure and volume overload with norepinephrine. *J. Mol. Cell. Cardiol.* **27,** 501–511.
34. Lee, W.M. (2003) Drug-induced hepatotoxicity. *N. Engl. J. Med.* **349,** 474–485.
35. Park, B.K., Kitteringham, N.R., Maggs, J.L., Pirmohamed, M., and Williams, D.P. (2005) The role of metabolic activation in drug-induced hepatotoxicity. *Annu. Rev. Pharmacol. Toxicol.* **45,** 177–202.
36. Kew, M.C. (2000) Serum aminotransferase concentration as evidence of hepato-cellular damage. *Lancet* **355,** 591–592.
37. Kaplan, M.M. (2002) Alanine aminotransferase levels: what's normal? *Ann. Intern. Med.* **137,** 49–51.
38. Bauer, I., Rensing, H., Florax, A., Ulrich, C., Pistorius, G., Redl, H., and Bauer, M. (2003) Expression pattern and regulation of heme oxygenase-1/heat shock protein 32 in human liver cells. *Shock* **20,** 116–122.
39. Fang, C.W., Yao, Y.M., Shi, Z.G., Yu, Y., Wu, Y., Lu, L.R., and Sheng, Z.Y. (2002) Lipopolysaccharide-binding protein and lipopolysaccharide receptor CD14 gene expression after thermal injury and its potential mechanism(s). *J. Trauma* **53,** 957–967.
40. Flo, T.H., Smith, K.D., Sato, S., Rodriguez, D.J., Holmes, M.A., Strong, R.K., et al. (2004) Lipocalin 2 mediates an innate immune response to bacterial infection by sequestrating iron. *Nature* **432,** 917–921.
41. Petricoin, E.F., Zoon, K.C., Kohn, E.C., Barrett, J.C., and Liotta, L.A. (2002) Clinical proteomics: translating benchside promise into bedside reality. *Nat. Rev. Drug Discov.* **1,** 683–695.
42. Fan, H. and Hegde, P.S. (2005) The transcriptome in blood: challenges and solutions for robust expression profiling. *Curr. Mol. Med.* **5,** 3–10.
43. Affymetrix Technical Note. (2003) An Analysis of Blood Processing Methods to Prepare Samples for GeneChip® Expression Profiling Santa Clara, CA.

10

From Pharmacogenomics to Translational Biomarkers

Donna L. Mendrick

Summary

There is a need for new biomarkers to enable faster detection of adverse events due to drugs and disease processes. One would prefer biomarkers that are useful in multiple species (i.e., translational or bridging biomarkers) so that it would be possible to directly link responses between species and follow such injury in both preclinical and clinical settings. This chapter will explore some of the issues surrounding the use of pharmacogenomics to identify and qualify such biomarkers, and examples will be provided.

Key Words: biomarkers; pharmacogenomics; toxicogenomics; translational.

1. Introduction

The term *toxicogenomics* is not used in regulatory language. The Food and Drug Administration (FDA) guidance document on the use of pharmacogenomics uses the term *pharmacogenomics* as an encompassing term for pharmacogenomics, pharmacogenetics, and toxicogenomics (http://www.fda.gov/ cder/genomics/default.htm). This led to some confusion, and an International Conference on Harmonisation (ICH) has released a draft guidance document on "Terminology in Pharmacogenomics" (http://www.fda.gov/cder/guidance/7619dft.pdf). Within this document, the term *pharmacogenomics* is defined as "the investigation of variations of DNA and RNA characteristics as related to drug response." Note that this term covers what has been called toxicogenomics at least as it affects drugs. *Pharmacogenetics* is defined as "a subset of pharmacogenomics" that examines "the influence of variations in DNA sequence on drug response." Because regulatory

From: *Methods in Molecular Biology, vol. 460: Essential Concepts in Toxicogenomics*
Edited by: D. L. Mendrick and W. B. Mattes © Humana Press, Totowa, NJ

language does not use the term *toxicogenomics* and this chapter is devoted to the comparison of gene expression changes among animals and humans, the term *pharmacogenomics* will be used herein. To complete the definitions used in this chapter, the word *translational* is used to reflect the need for biomarkers that bridge across species that are used in preclinical settings and in humans.

Pharmacogenomics has been embraced by many in the pharmaceutical and regulatory arenas. An advisory group to the secretary of the Department of Health and Human Services recently issued a draft report on its promises and challenges. Among the promises listed is the ability to address major health needs including decreasing adverse drug reactions (ADRs) and translating preclinical studies to clinical trials and finally to clinical practice (www4.od.nih.gov/oba/sacghs/reports/CR_report.pdf). Unfortunately, there is no agreed upon pathway to qualify biomarkers and such is needed to spur the field forward. This chapter will address the use of pharmacogenomics to identify and begin to qualify translational biomarkers and its associated issues.

2. Classic Approaches

At a November 2006 meeting cosponsored by the FDA on "Best Practices and Development of Standards for the Submission of Genomic Data to the FDA," Felix Frueh (Associate Director, Genomics, Office of Clinical Pharmacology, Office of Translational Science, CDER, FDA) stated that most biomarkers in use today were found after decades of research and discussion and many were not fully assessed using the type of biological and analytical validation required today. Currently used markers of renal dysfunction, serum creatinine and blood urea nitrogen (BUN), are relatively insensitive and do not provide sufficient notice of renal injury to take preventative steps particularly in the area of acute injury *(1)*. Kidney injury molecule 1 (Kim-1) was identified a decade ago as a potential biomarker of renal proximal tubule injury using an animal model of human acute renal injury *(2)*. Many researchers have studied this gene and its encoded protein and found it upregulated upon various types of injuries including toxicant exposure and disease processes as will be discussed in more detail later. Even though this biomarker addresses an unmet clinical need, was proposed a decade ago, and multiple researchers and a consortium group have worked on it, only in summer 2007 will it be officially proposed as a biomarker of renal injury in preclinical species *(1–10)*. (Still remaining is its qualification in the clinical arena.) The need to improve our ability to avoid toxic drugs and chemicals and to embrace personalized medicine means that a more streamlined process of biomarker qualification must be adopted.

3. Pharmacogenomics

3.1. Translational Biomarkers in Common Usage

As noted previously, the term *toxicogenomics* is not used in regulatory language and, in fact, can be misleading as it is known that many genes respond in a common manner to disease processes and toxicants leading some to mistakenly believe that indicators of toxicity and disease cannot be the same. Common biomarkers such as alanine aminotransferase (ALT) denote hepatocellular injury regardless of its origin and are used in both preclinical and clinical settings. Although one might wish for a single biomarker for ease in interpretation, multiple biomarkers are in use today in both preclinical and clinical settings and likely the same will be true for new biomarkers. As an example, liver dysfunction and injury is monitored through the use of assays that monitor serum levels of ALT, aspartate aminotransferase (AST), alkaline phosphatase, bilirubin, cholesterol, and so forth. No one marker is deemed to be of sufficient specificity and sensitivity. The FDA realized that a panel of biomarkers likely are required and stated such in their pharmacogenomics guidance document, yet some individuals still seem to hold this new field to unattainable standards in terms of demanding an assay use a single biomarker and perform at unachievable levels of accuracy. Hopefully, this will abate as more individuals become comfortable with the science. It is important to keep in mind that a large number of drugs that prove to have severe adverse events enter the clinic as classic testing cannot detect them *(11)*. Therefore, any new science should be compared with existing accuracy instead of an ethereal standard.

3.2. Biomarker Purpose

Before beginning any scientific experiment, one must design the study appropriately to answer the questions being posed. In this case, one prime question is: what is the purpose of the proposed biomarker? If it is to predict that a tissue injury will occur, then one needs to focus on genes whose expression level changes prior to the phenotypic injury whether or not the expression level of the genes remain altered at the time of observable injury. For the purposes of this chapter, such genes and their encoded proteins will be called predictive biomarkers. Alternatively, one might want to find biomarkers that diagnose tissue injury. In this instance, the focus would be on gene expression changes at the time of observable tissue injury. These can be termed diagnostic biomarkers.

Once one determines the purpose of the biomarker's eventual use, it is important to consider the sampling method that will need to be employed in the clinic. A diagnostic biomarker that provides differential information regarding tumors might be useful as an RNA biomarker because that tissue is generally removed for study. However, a diagnostic biomarker of solid tissue injury might be most useful if one did not have to depend on biopsies as the cost of biopsies is large and this invasive test can cause morbidity and mortality *(12)*. Predictive biomarkers are meant to identify subtle injury prior to observable damage so generating an assay that depends on tissue RNA would not be useful in the clinic. Therefore, the options for such biomarkers are to format an assay that can monitor protein changes in body fluids or follow gene expression changes in circulating white blood cells. An interesting example of a gene expression test showing clinical utility in place of biopsies is an assay that uses 20 genes on circulating peripheral blood mononuclear cells (PBMCs) to identify patients undergoing cardiac allograft rejection. During the qualification of this test, it was recognized that the "gold standard" of pathologists' assessment of rejection was hampered by observer differences, and it is hypothesized that this quantitative test might remove some of this variability. This test is used to identify patients undergoing rejection and is now under evaluation to determine its ability to predict future acute cellular rejection *(13)*. There are some examples in the literature where researchers have found gene expression changes in circulating white blood cells that indicate organ dysfunction or injury *(14–16)*. Examples in the field of toxicology include a paper by McHale et al. on the detection of individuals exposed to dioxin more than 20 years ago *(17)* and the work being done in Rick Paules' group at the National Institute of Environmental Health Sciences (NIEHS) where they are comparing the responses of white blood cells in rats and humans, both exposed *in vivo* to acetaminophen *(18)*. However, much still remains to be examined in this sentinel approach particularly as few of these biomarkers have yet to be qualified. For example, will gene expression changes in circulating white blood cells enable detection and localization of the dysfunction (e.g., liver versus kidney) or solely enable the clinician to realize there is toxicity occurring somewhere in the body?

3.3. Pharmacogenomic Data Analysis Leading to New Biomarkers

As discussed above, clinical use would be best served if biomarkers can be monitored in body fluids or on circulating white blood cells. One can begin with the examination of gene expression changes and focus the analysis on those genes known to encode secreted proteins or those on the membrane of cells that might be cleaved upon injury. In the latter instance, extracellular domains of membrane proteins might appear in body fluids. There are published examples of a pharmacogenomic approach leading to the identification of protein biomarkers: the first is a secreted protein and the second is as membrane bound protein elicited by injury and then cleaved from the cell surface.

3.3.1. Adipsin

Proprietary drugs have been developed that inhibit γ-secretase as a potential therapy for Alzheimer's disease. However, many have been shown to inhibit Notch processing, which plays a role in replenishing the gastrointestinal tract and immune system in adults. Two reports have found a link between such drugs and gastrointestinal toxicity; the first was from a group at Eli Lilly and Company and the second from researchers at AstraZeneca Pharmaceuticals *(19,20)*. The former group examined the response to their proprietary compound X. Treatment of Fischer 344 rats with this compound for 4 days induced a gastrointestinal toxicity that was exhibited by, among other signs, increased tissue weight and number of goblet cells with a reduced number of enterocytes. Gene expression studies found adipsin, a gene encoding a serine protease, was upregulated in the affected tissue. These investigators also found the expression level of this gene was increased by exposure of rat and human intestinal epithelial-derived cell lines to compound X thus showing conservation of this response between *in vivo* and *in vitro* environments and between species. They examined the intestinal contents after rat exposure to compound X and found adipsin protein levels were increased *(19)*. Scientists at AstraZeneca examined the response of Han Wistar rats to exposure to three of their proprietary γ-secretase inhibitors. They found genes altered early in the time course that may serve as early predictive biomarkers, and this set included adipsin. They confirmed an increased presence of adipsin protein in the feces of treated rats *(20)*. Adipsin is also known as complement factor D, C3 convertase activator, and properdin factor D as noted by SwissProt (http://au.expasy.org). Together these studies illustrate that adipsin may prove to be a translational biomarker of advancing gastrointestinal toxicity that can be monitored by examining the level of gene expression in biopsied intestine or as a protein in feces. The gene encoding this protein is present in many normal tissues of the body as shown in **Fig. 1** suggesting it might not be a good biomarker of tissue injury if monitored in the blood. Interestingly, the concept of monitoring the protein in intestinal contents might render more specificity to gastrointestinal damage although this remains to be proved.

3.3.2. Kim-1/HAVCR1

There are deficiencies with current serum markers of renal injury as noted above and, in some instances such as acute renal failure, therapeutic intervention may be too late by the time such markers are altered. Thus, there is an impetus

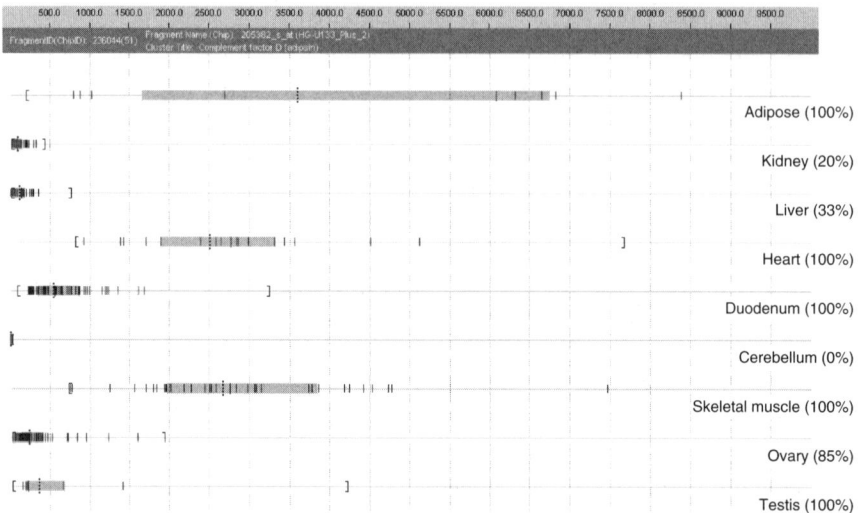

Fig. 1. e-Northern™ analysis of adipsin expression in normal human tissues. As visualized using the Genesis Enterprise System® software (Gene Logic, Gaithersburg, MD), the signal intensity (linear scale) of each biological sample for adipsin, as represented by fragment 205382_s_at on the Affymetrix GeneChip® Human Genome U133 Plus 2.0 Array, is shown on the x axis, and the respective tissues are listed on the y axis. The scatterplot graphs each individual sample type as a line. The percentage of samples where the gene is called present is shown in the parentheses after each group name. The dotted line represents the mean of the group, and the encompassing hatch marks is the interquartile range. As can be seen, adipsin is expressed by most tissues examined.

to advance the discovery and qualification of new biomarkers. As with all solid tissue, one can monitor the tissue directly for either gene expression changes or protein abnormalities as well as examine the blood for changes in secreted proteins levels, and so forth. For all types of tissue injury, it is possible to monitor the level of proteins in the urine although that would require the protein to be of a small enough size and the correct molecular charge to pass through the glomerular filtration apparatus and not be reabsorbed by the tubules as part of their normal functioning mode. If biomarkers of renal tubular damage are sought, it is more likely than with other tissues that one can find proteins shed or secreted directly into the urine without concerns about passing through the glomerular filtration system. It is still important to determine the tissue specificity of the biomarker as secretion from other tissues may cloud the biomarker's use particularly if the patients or animals have renal injury whereby the protein may have abnormal access to the urine.

One promising biomarker of renal tubular injury is kidney injury molecule-1, or Kim-1. The rat version of this gene is known as Kim-1. The human ortholog is called HAVCR1 (hepatitis A virus cellular receptor 1), and the murine version is named Havcr1. It also is known as T-cell immunoglobulin and mucin domain–containing protein 1 (TIMD-1) or T-cell membrane protein 1 (TIM-1). In all species, the protein it encodes is known as HAVcr-1. To avoid confusion in this discussion, the term *Kim-1* will be used herein for the gene and protein of all species. The potential for this gene or its encoded protein to be a useful renal proximal tubule biomarker was discovered by Bonventre et al. in 1998. These investigators were working with a model of renal ischemia in rodents as a counterpart to the ischemia that can occur in humans and can lead to acute kidney injury. Using a genomic approach, they found Kim-1 was dramatically expressed in the ischemia rat kidney but at very low levels in the nonischemic contralateral kidney. They cloned the human counterpart and discovered the similarity to HAVCR1. Using *in situ* hybridization and immunohistochemistry, they reported that the gene and protein, respectively, were expressed in proliferating and dedifferentiated epithelial cells in regenerating proximal tubules *(2)*. Bonventre and his colleagues developed monoclonal antibodies to the human version of Kim-1, and immunohistochemistry revealed increased protein levels in biopsies of six patients with acute tubular necrosis. Kim-1 is a membrane bound protein whose ectodomain is shed upon injury to the cell. Using an immunoassay, they reported an elevated level of Kim-1 protein in the urine of patients with ischemic acute tubular necrosis *(6)*. Bonventre and others have begun to explore the expression and excretion of Kim-1 in other renal diseases. Polycystic kidney disease (PKD) is caused predominately by an inherited defect that is thought to lead to dedifferentiation of renal proximal tubule cells and cyst formation. Because Kim-1 was found to be associated with such tubular simplification in renal ischemia, Bonventre and his co-workers examined a murine model of PKD and found increased expression of Kim-1 protein in affected tubules *(8)*. The increased presence of Kim-1 protein in renal cell carcinoma was recently described *(21)*. Examination of Kim-1 in human renal diseases was performed using the BioExpress® database at Gene Logic (Gaithersburg, MD). The expression of Kim-1 mRNA was significantly increased, compared with normal human kidney, in patients with PKD, glomerulosclerosis (GS), and primary malignancies including clear cell adenocarcinoma, renal cell carcinoma, and transitional cell carcinoma. Interestingly, patients with a benign neoplasm of the kidney exhibited a significantly reduced expression of Kim-1 suggesting the potential for Kim-1 to be useful in differential diagnosis (**Fig. 2**).

To perform well as a protein biomarker, it is best if the tissue localization is specific to the kidney. We previously have reported that Kim-1 is expressed

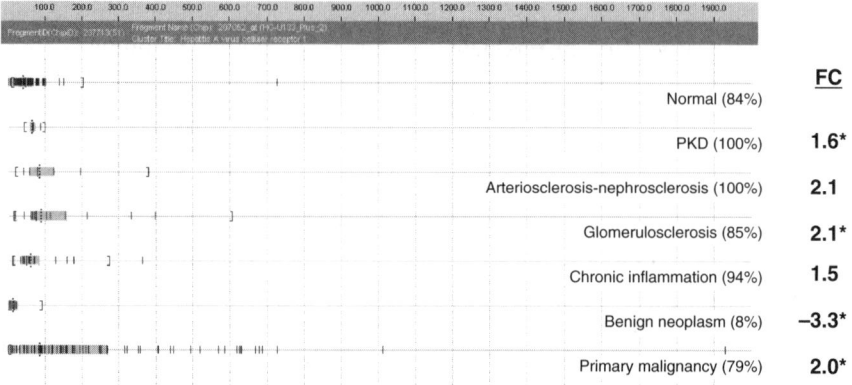

Fig. 2. e-Northern analysis of Kim-1 (IIAVCR1) in human renal diseases. (As noted in the text, the term KIM-1 will be used for all orthologs to conform to current vernacular.) Kim-1 expression was assessed by evaluating the expression level of fragment 207052_at on the Affymetrix GeneChip Human Genome U133 Plus 2.0 Array. Fold change (FC) was calculated by comparing diseased tissue to normal human kidney, and statistically significant changes ($p < .05$) are shown with an asterisk. Kim-1 expression is significantly increased in PKD, GS, and malignant tumors of the kidney but is decreased in benign tumors. For details on the e-Northern visual, please see the **Fig. 1** legend.

predominately by the kidney but not other tissues in the wide panel of normal human tissues examined *(22)*. The fact that Kim-1 was not found in normal thymus and white cells may be surprising because it also is known as T-cell membrane protein-1 (TIM-1). This may be explained by the fact that this gene has been shown to be expressed on activated T cells and is either not present on naïve T cells or found at very low levels *(23–25)*. To be useful as a translational protein biomarker, this gene should not be expressed by many normal tissues of the rat and mouse, and that is the case (**Fig. 3** and Ref. *21*, respectively). Many investigators have reported an increase in Kim-1 expression after injury to the proximal tubule caused by drugs and toxicants in rats and primates, and most have reported the gene and protein to be altered at the time of injury, which can be observed by histopathology *(1,3,5,26)*.

Studies from the ToxExpress database were selected to examine the time course of gene expression change and the results from three illustrative studies are shown in **Figs 4 to 6**. Cisplatin induces injury first to the proximal tubule and then the damage spreads to other tubule segments, whereas lithium first attacks the distal tubule causing, in acute toxicity, tubular necrosis, and chronic dosing can damage distal tubules, collecting ducts, and the tubulointerstitium *(27–29)*. Dantrolene's drug label reflects its potential to induce occlusion

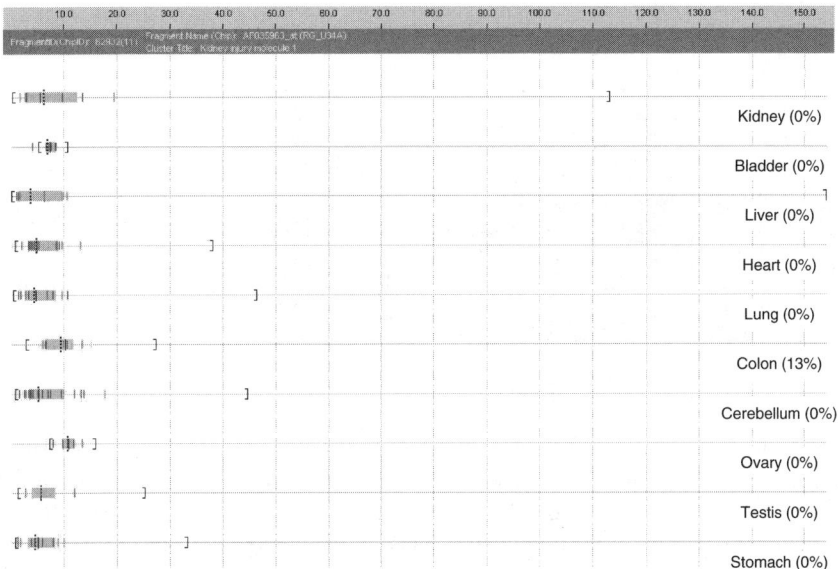

Fig. 3. e-Northern analysis of Kim-1 in normal rat tissues. Kim-1 is represented by fragment AF035963_at on the Affymetrix GeneChip Rat Genome U34A Array. It is not expressed to any extent in any normal rat tissue examined. For details on the e-Northern visual, please see the **Fig. 1** legend.

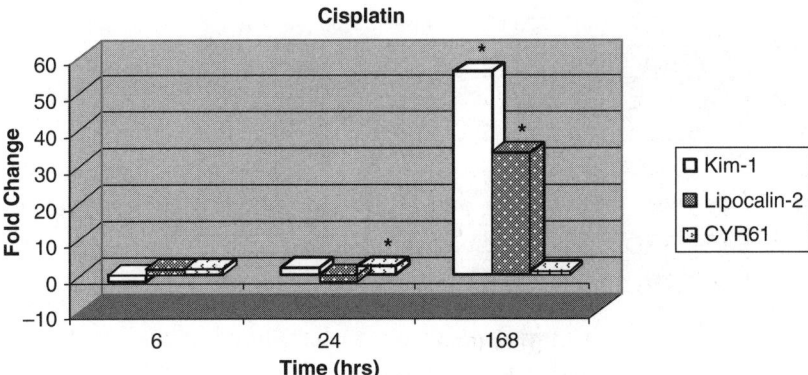

Fig. 4. Response of renal levels of Kim-1, lipocalin-2, and CYR61 to cisplatin treatment. The graph illustrates the fold change of the mean of the high-dose cisplatin-treated group versus the corresponding vehicle group sacrificed at the same time point. The treatment conditions and time course wherein the change in gene expression from normal yields a statistical significance of $p < .05$ is indicated by an asterisk. Kim-1 was significantly elevated at 168 h after dosing as was lipocalin-2. CYR61 expression was increased at 24 h after dosing.

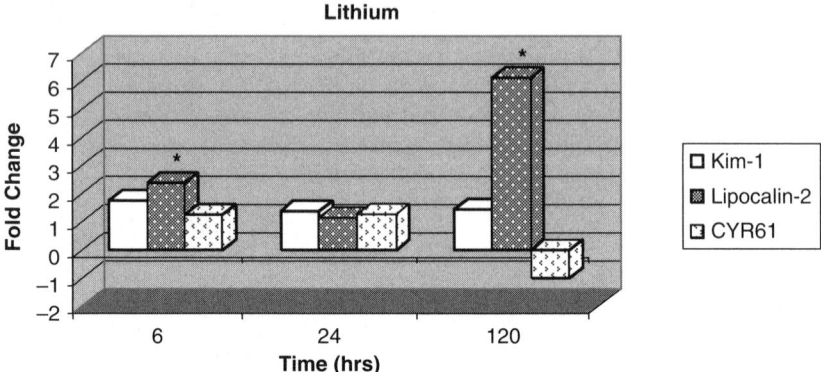

Fig. 5. Response of renal levels of Kim-1, lipocalin-2, and CYR61 to lithium treatment. The graph illustrates the fold change of the mean of the high-dose lithium-treated group versus the corresponding vehicle group sacrificed at the same time point. The treatment conditions and time course wherein the change in gene expression from normal yields a statistical significance of $p < .05$ is indicated by an asterisk. Lipocalin-2 expression was elevated at 6 and 120 h, whereas no significant changes were seen in Kim-1 or CYR61 gene expression.

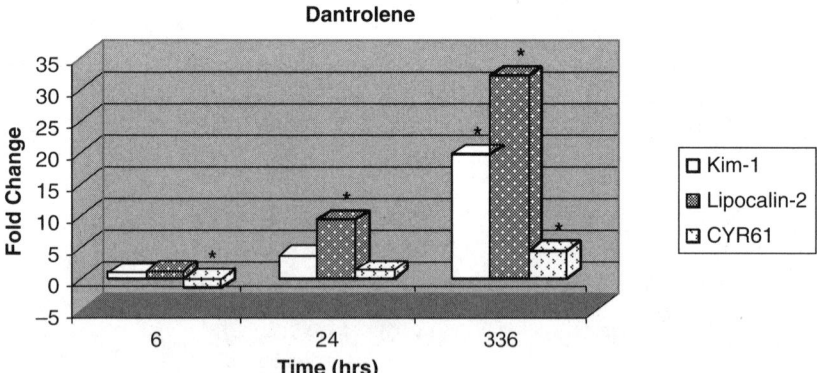

Fig. 6. Response of renal levels of Kim-1, lipocalin-2 and CYR61 to dantrolene treatment. The graph illustrates the fold change of the mean of the high-dose dantrolene-treated group versus the corresponding vehicle group sacrificed at the same time point. The treatment conditions and time course wherein the change in gene expression from normal yields a statistical significance of $p < .05$ is indicated by an asterisk. Kim-1 expression was elevated at 336 h. Lipocalin-2 gene expression was increased at 24 and 336 h after exposure. CYR61 levels were lowered at 6 h and increased at 336 h after dose.

nephropathy. For these three studies, male Sprague-Dawley (SD) rats were used. Rats were injected intravenously once with vehicle, 1 or 5 mg/kg body weight of cisplatin, and groups sacrificed 6, 24, or 168 h post dose. Clinical pathology and histopathology were performed at all time points and urinalysis at 24 h and 168 h post dose. Only at 168 h were histologic changes and urinalysis abnormalities noted. There was renal tubule cells involvement, especially the distal convoluted tubules, with evidence of degeneration, necrosis, and regeneration of tubules. Lithium was administered by intraperitoneal injection daily at doses of 0, 12.4, and 124.4 mg/kg body weight per day. Groups of rats were sacrificed at 6 and 24 h after one treatment and 120 h after daily administration. At the high dose, it induced a nephrosis evident only at 120 h. It appeared to be limited to the cortical connecting tubules and to the connecting ducts of the cortex and outer medulla. Serum creatinine was elevated progressively at 6 and 120 h. Dantrolene was administered as daily gavage at 0, 50, and 2000 mg/kg body weight per day and rats sacrificed 6 and 24 h after the first dose and at 336 h after daily dosing. Rats treated with the high dose of dantrolene exhibited an increase in serum creatinine at 24 and 336 h and in proteinuria at 336 h. Histologic examination revealed that rats exposed to high-dose dantrolene for 336 h experienced nephrosis that involved all regions of the kidney and was typically associated with multifocal acute inflammatory infiltrates.

As can be seen in **Figs. 4 to 6**, Kim-1 gene expression is significantly elevated only at the time of injury, which can be observed with histopathology. However, because renal biopsies are invasive and expensive, Kim-1 can serve as an important biomarker in both preclinical and clinical arenas. Novartis is working with the FDA through a CRADA and with the C-Path Institute to qualify renal biomarkers, including Kim-1. It is unfortunate that this process has taken almost 10 years from the time Kim-1 was discovered.

3.3.3. Lipocalin-2/LCN2

Lipocalin was originally identified as a protein bound to gelatinase from human neutrophils and thus is also known as neutrophil gelatinase-associated lipocalin (NGAL). The official gene symbol is LCN2 for human and Lcn2 for rats and mice. NGAL and lipocalin-2 are used to refer to the protein (http://au.expasy.org). For ease in discussion, the term *lipocalin-2* will be used here for both gene and protein. Lipocalin-2 is a secreted protein that is expressed by many normal and diseased human tissues including lung, stomach, small intestine, pancreas, kidney, thymus, and adenocarcinomas of the lung, colon, and pancreas. In contrast, renal cell carcinomas express low levels of this protein *(30)*. Such a pattern of gene expression is observed as well in normal human tissues (**Fig. 7**) and in malignant human tissues (**Fig. 8**).

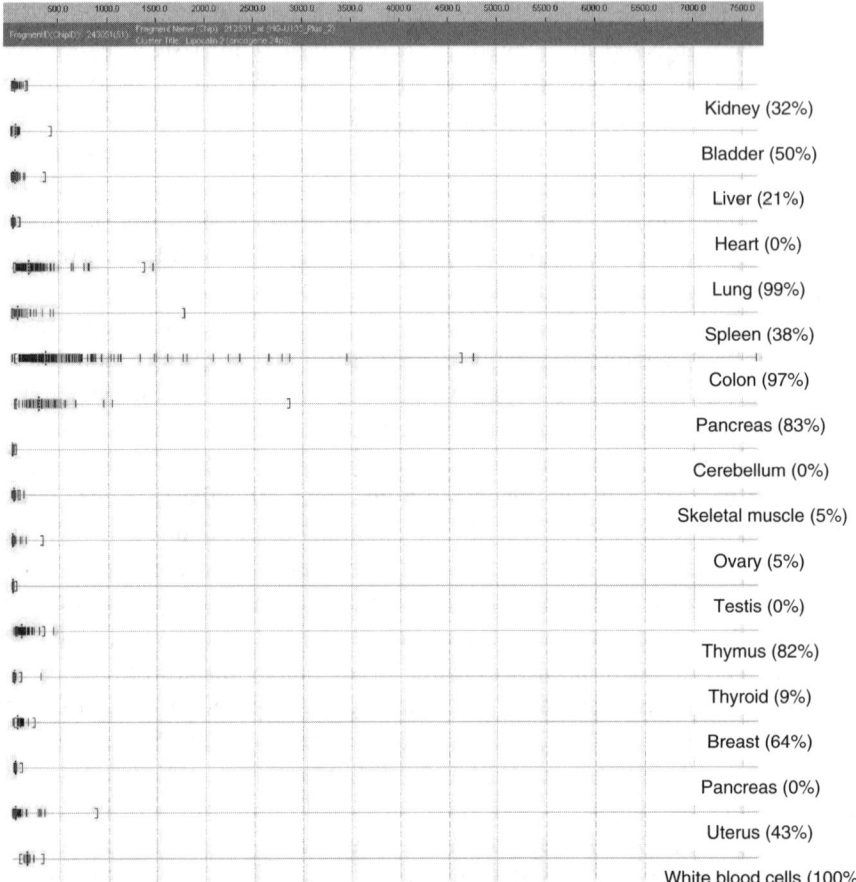

Fig. 7. e-Northern analysis of lipocalin-2 in normal human tissue. Lipocalin is represented by fragment 212531_at on the Affymetrix GeneChip Human Genome U133 Plus 2.0 Array. It is expressed by the majority of the normal lung, colon, pancreas, thymus, breast, and white blood cell samples. It is expressed by 32% of the normal renal samples surveyed but at a very low signal level. For details on the e-Northern visual, please see the **Fig. 1** legend.

Figures 9 to 11 illustrate that the expression of lipocalin-2 is found in many normal tissues of the rat, canine, and mouse, and multitissue localization has been reported by others *(30)*. In 2003, Mishra and colleagues *(31)* used microarray technology and found that lipocalin-2 was upregulated within 3 h in postischemic mouse kidneys compared with the contralateral kidney, and this was confirmed by RT-PCR. These same authors examined the protein level in the kidneys using Western analysis and reported an increase in protein, also

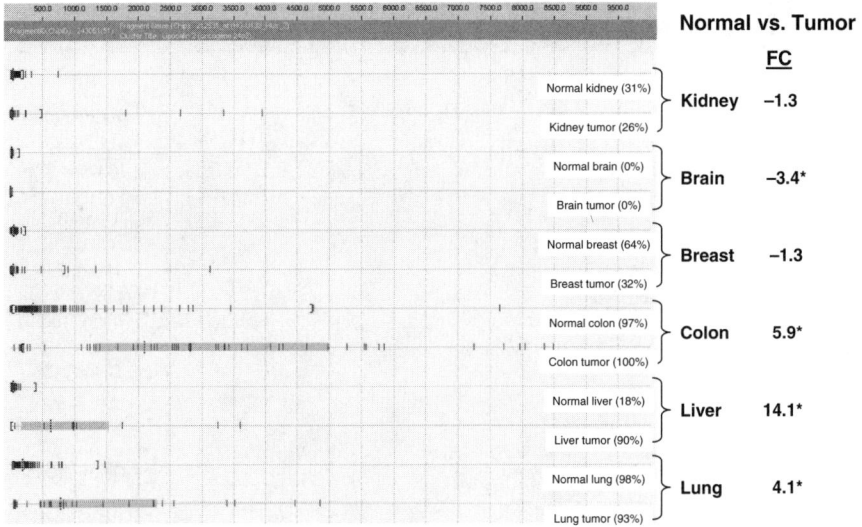

Fig. 8. e-Northern analysis of lipocalin-2 in malignant tissues. The fold change (FC) was calculated as the expression level of the malignant tissue compared with the normal tissue. Statistical significance is defined as $p < .05$ and illustrated by an asterisk. Lipocalin-2 is significantly increased in tumors of the colon (5.9-fold), liver (14.1-fold), and lung (4.1-fold) but is not altered in renal tumors. For details on the e-Northern visual, please see the **Fig. 1** legend.

seen within 3 h of injury. Although lipocalin-2 protein was not found in normal mouse urine, it was present within 2 h of injury; a time far earlier than N-acetyl-β-D-glucosamine (NAG), a commonly used biomarker of renal injury. They also reported an increase in lipocalin-2 protein concentration in rat urine within 3 h of ischemic injury and within 1 day of the administration of nephrotoxic doses of cisplatin *(31)*. In our studies, lipocalin-2 gene expression was significantly increased only at 168 h after cisplatin treatment (**Fig. 4**), the time of observable injury, but was increased at 6 h and 120 h after lithium exposure (**Fig. 5**) and at 24 h and 336 h after dantrolene administration (**Fig. 6**) suggesting the gene may be a good predictive marker of incipient renal injury. In humans, the serum level of lipocalin-2 is increased in patients with bacterial infections and renal tubular injury and in the urine of patients with renal tubular injury and breast cancer *(1,32)*. Examination of diseased human kidneys reveals the lipocalin-2 gene upregulated in PKD, nephrosclerosis, and GS, as well as in chronic inflammation (**Fig. 12**). It was not altered by renal malignancies or benign neoplasms suggesting it may offer some differential diagnostic potential. All of these reports suggest that lipocalin-2 might be a good translational biomarker of renal injury and, in fact, may serve as a predictive biomarker of

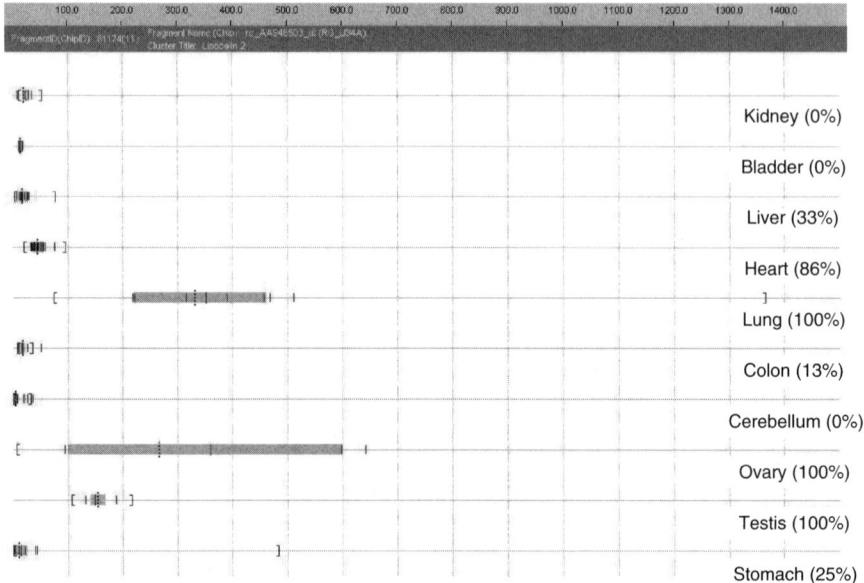

Fig. 9. e-Northern analysis of lipocalin-2 in normal rat tissues. Lipocalin-2 is represented by fragment rc_AA946503_at on the Affymetrix GeneChip Rat Genome U34A Array. It is expressed, at varying levels, in the majority of samples from the normal heart, lung, ovary, and testis but not in normal rat kidney. For details on the e-Northern visual, please see the **Fig. 1** legend.

renal damage. However, the ubiquitous expression of the lipocalin-2 gene and protein and reports of elevated protein levels in serum and urine in nonrenal diseases suggest specificity may be an issue if this biomarker were used alone.

3.3.4. Cysteine-Rich Protein 61/CYR61

CYR61 is the name of the gene in humans with Cyr61 the symbol for the mouse and rat orthologs. CYR61 is the name of the protein in these same species (http://au.expasy.org). To avoid confusion, the term *CYR61* will be used for all. CYR61 is a secreted protein that binds several integrins, is associated with the extracellular matrix membrane, and regulates angiogenesis. As such, it has been associated with wound healing, tumorigenesis, fibrosis, and vascular disorders. It is transiently expressed in regenerating liver *(33)*. Elevated levels of CYR61 protein in breast cancer are associated with more advanced disease, and it has been proposed that its interactions with $\alpha_v\beta_3$ helps protect breast cancer cells from chemotherapy *(34)*. Its potential as a biomarker of renal injury arose from a study done by Muramatsu et al. in 2002 *(35)*.

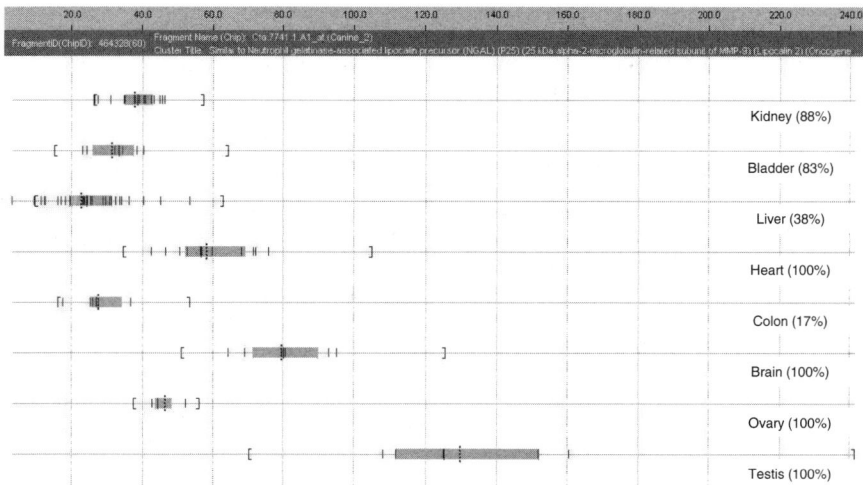

Fig. 10. e-Northern analysis of lipocalin-2 expression in normal canine tissues. Lipocalin-2 expression was assessed by evaluating expression levels of fragment Cfa.7741.1.A1_at on the Affymetrix GeneChip Canine Genome 2.0 Array. Lipocalin-2 is expressed by the majority of normal kidney, bladder, heart, brain, and ovary samples with the highest expression in the heart, brain, and testis. For details on the e-Northern visual, please see the **Fig. 1** legend.

A genomic approach was applied to models of renal ischemia, and CYR61 was one of the genes identified whose expression levels dramatically increased. Using Northern blot analysis on normal rat tissues, these authors reported its presence in heart, lung, kidney, and skeletal muscle but not in spleen or liver. Western blot analysis revealed an increase in CYR61 protein in the affected kidney within 1 h of ischemia. Localization of CYR61 mRNA was performed with *in situ* hybridization and, although it was not detectable in normal rat kidney, it was found in the proximal straight tubules 2 h after bilateral ischemia. CYR61 protein was not found in normal rat urine but was detectable 3 to 6 h after ischemia *(35)*. However, its usefulness as a urinary biomarker of renal injury has been questioned because the levels decrease over time even though the kidney is undergoing progressive injury *(1)*. Examination of our selected studies revealed that CYR61 gene expression was significantly increased at 24 h after cisplatin administration (**Fig. 4**) but was not affected by lithium treatment (**Fig. 5**). Exposure to dantrolene caused the gene to be downregulated at 6 h and increased at 336 h (**Fig. 6**). Together these results suggest a potential for CYR61 to be a predictive and diagnostic biomarker. The latter was further suggested by examination of human renal diseases wherein CYR61 mRNA levels were increased in PKD, GS, and chronic inflammation. Conversely,

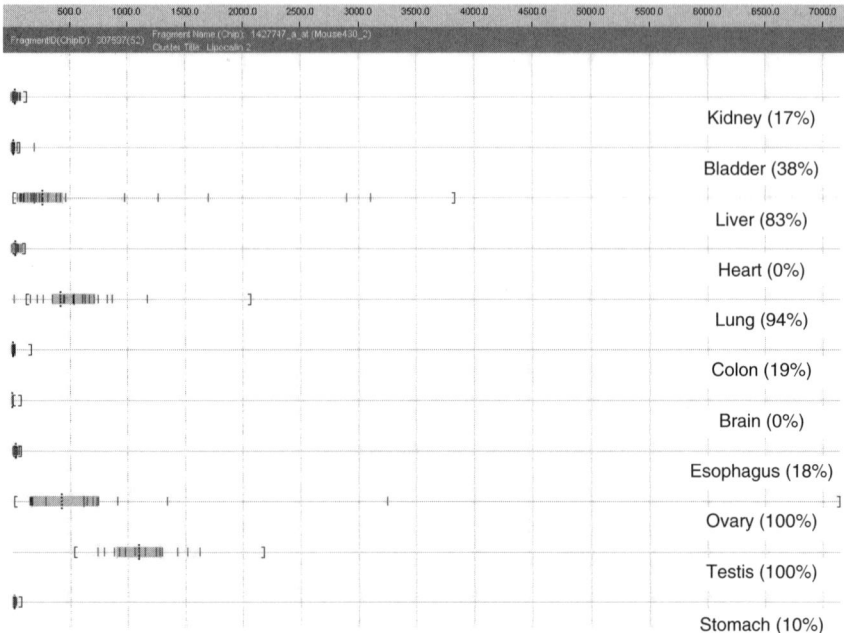

Fig. 11. e-Northern analysis of lipocalin-2 expression in normal mouse tissues. Lipocalin-2 is represented by fragment 1427747_at on the Affymetrix GeneChip Mouse Genome 430 2.0 Array. Lipocalin-2 is expressed by the majority of liver, lung, ovary, and testis samples surveyed. For details on the e-Northern visual, please see the **Fig. 1** legend.

levels were decreased in benign or malignant renal tumors suggesting some potential of this gene and protein to serve as differential diagnostic biomarkers (**Fig. 13**).

CYR61 was also downregulated in malignant tumors of the kidney, breast, colon, and lung as illustrated in **Fig. 14** suggesting a decrease in protein levels in body fluid might be a ubiquitous indicator of tissue malignancy.

A survey of normal tissues shows that this gene is ubiquitously expressed in normal human and murine tissues but less universally observed in normal rat and canine tissues (**Figs. 15 to 18**). This tissue distribution was more diffuse than previously reported as discussed above.

In summary, the ubiquitous nature of the expression of CYR61 among normal tissues as reported by others and seen in the data presented here suggests that CYR61 may serve as a predictive and diagnostic protein biomarker although it may not have the required specificity as a single biomarker. However, it may exert some power as a member of a biomarker panel.

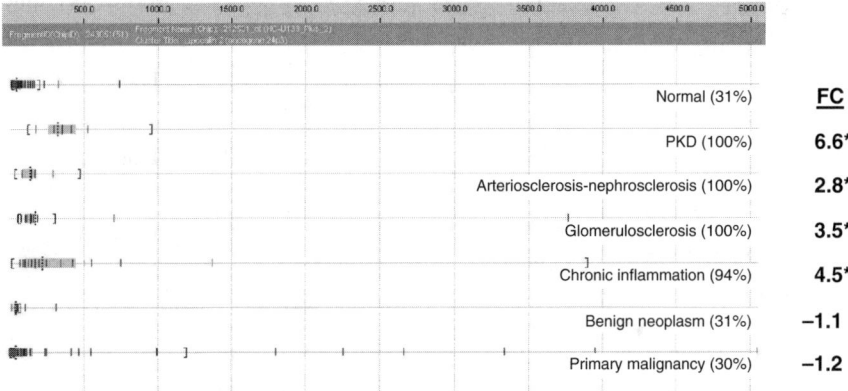

Fig. 12. e-Northern analysis of lipocalin-2 in human renal diseases. Fold change (FC) was calculated by comparing diseased tissue to normal human kidney, and statistically significant changes ($p < .05$) are shown with an asterisk. Lipocalin-2 expression is significantly elevated in PKD, nephrosclerosis, GS, and chronic inflammation but not in benign or malignant tumors. For details on the e-Northern visual, please see the **Fig. 1** legend.

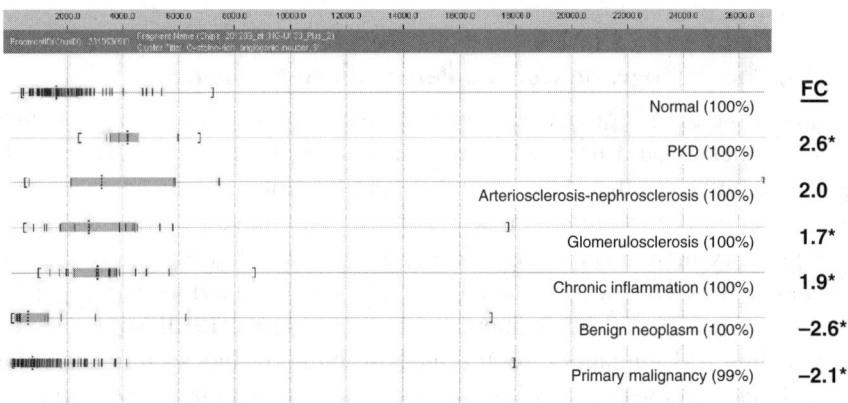

Fig. 13. e-Northern analysis of CYR61 in human renal diseases. CYR61 is represented by fragment 201289_at on the Affymetrix GeneChip Human Genome U133 Plus 2.0 Array. Fold change (FC) was calculated by comparing diseased tissue to normal human kidney, and statistically significant changes ($p < .05$) are shown with an asterisk. CYR61 expression is significantly elevated in PKD (2.6-fold), GS (1.7-fold), chronic inflammation (1.9-fold), and decreased in benign (–2.6-fold) and malignant (–2.1-fold) tumors. For details on the e-Northern visual, please see the **Fig. 1** legend.

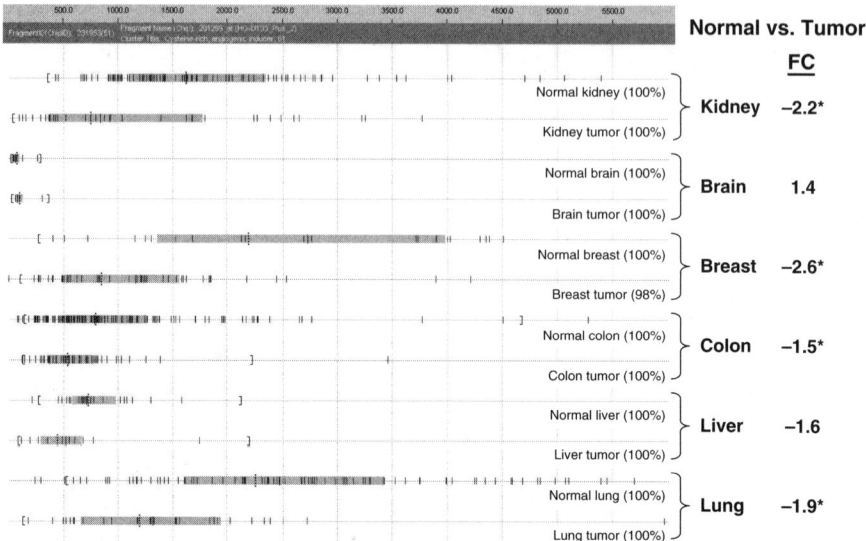

Fig. 14. e-Northern analysis of CYR61 in malignant tissues. The fold change (FC) was calculated as the expression level of the malignant tissue compared with the normal tissue. Statistical significance is defined as $p < .05$ and is illustrated with an asterisk. CYR61 is significantly decreased in renal, breast, colon, and lung tumors. For details on the e-Northern visual, please see the **Fig. 1** legend.

4. FDA, Pharmacogenomics, and Personalized Medicine

Without the FDA's proactive stance on pharmacogenomics and personalized medicine, there would be much less advancement in its use in understanding drugs' efficacy and safety leading to the identification of biomarkers. The FDA quickly recognized the impact that genomics could have in drug development and regulatory actions and cosponsored, with industry groups, a meeting that was held on May 16–17, 2002, which was attended by individuals from pharmaceutical companies and other interested parties. **Figure 19** illustrates some of the meetings and guidances, represented by building blocks in a house, that the FDA has been using to enable pharmacogenomics to be utilized to deliver qualified and analytically validated biomarkers.

At this first meeting, the FDA used the term *safe harbor* to describe how exploratory genomic data might be submitted to the agency without regulatory impact until more is understood about the appropriate interpretation of such data. This began the proactive process of the FDA encouraging companies to explore genomics as a means to find biomarkers, whether they are gene expression–based or SNP (single nucleotide polymorphism)-based, to spur drug development without fear of regulatory reprisals *(22)*.

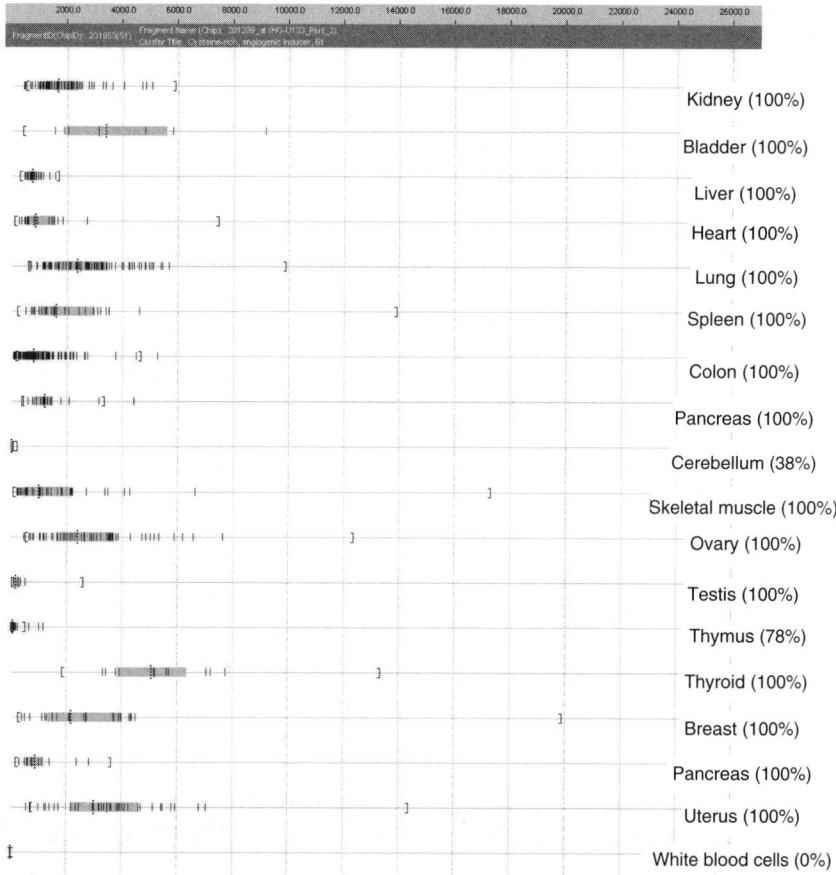

Fig. 15. e-Northern Analysis of CYR61 in normal human tissues. CYR61 is ubiquitously expressed in normal human tissues. For details on the e-Northern visual, please see the **Fig. 1** legend.

In the fall 2003, the FDA issued a draft guidance document on the use of pharmacogenomics, and the second FDA-sponsored workshop quickly followed. The discussions at this second workshop clearly illustrated the industry's interest in pursuing pharmacogenomics, yet the draft guidance contained ambiguities in both the preclinical and clinical use of genomic data as was reflected in some of the papers that followed the meeting *(36–39)*. At approximately the same time this was happening, the FDA prepared a white paper commonly named the Critical Path Document (http://www.fda.gov/oc/initiatives/criticalpath/whitepaper.html) wherein the agency stated a need for an effort to apply new biomedical science to product

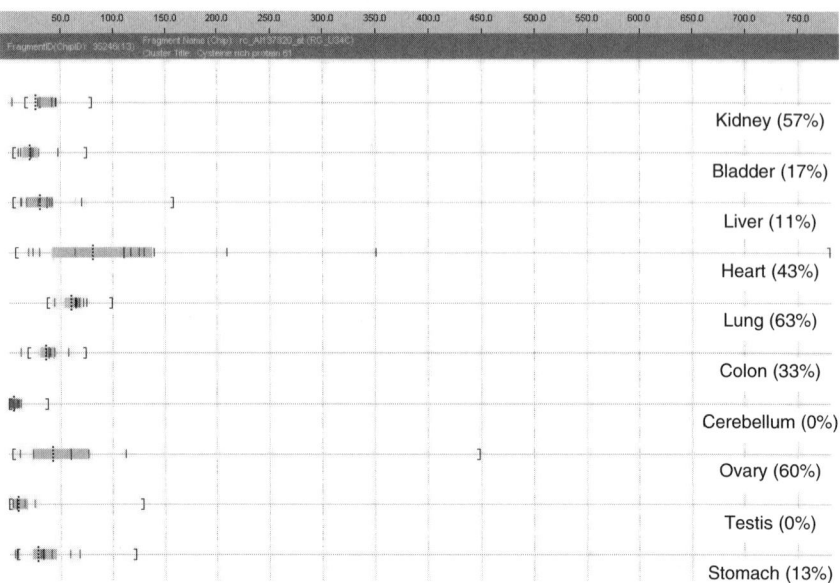

Fig. 16. e-Northern Analysis of CYR61 in normal rat tissues. CYR61is represented by fragment rc_AI137820_at on the Affymetrix GeneChip Rat Genome U34A Array. There is limited expression in normal kidney, heart, lung, colon, and ovary. For details on the e-Northern visual, please see the **Fig. 1** legend.

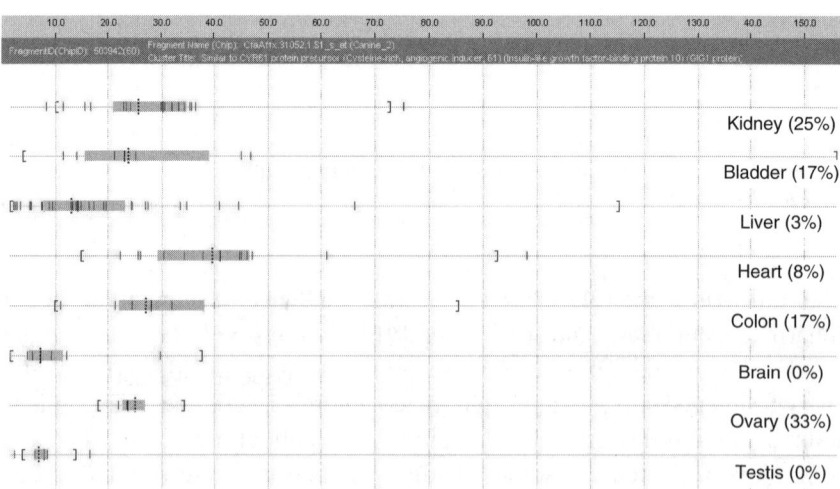

Fig. 17. e-Northern analysis of CYR61 in normal canine tissue. CYR61 expression was assessed by evaluating expression levels of fragment CfaAffx.31052.1.S1_s_at on the Affymetrix GeneChip Canine Genome 2.0 Array. There was very limited expression in the normal tissues surveyed. For details on the e-Northern visual, please see the **Fig. 1** legend.

50.0
100.0
150.0
200.0
250.0
300.0
350.0
400.0
450.0
500.0
550.0
600.0
650.0
700.0
750.0
800.0
850.0
Kidney (100%)
Bladder (100%)
Liver (25%)
Heart (100%)
Lung (94%)
Colon (57%)
Brain (25%)
Esophagus (82%)
Ovary (30%)
Testis (0%)
Stomach (70%)

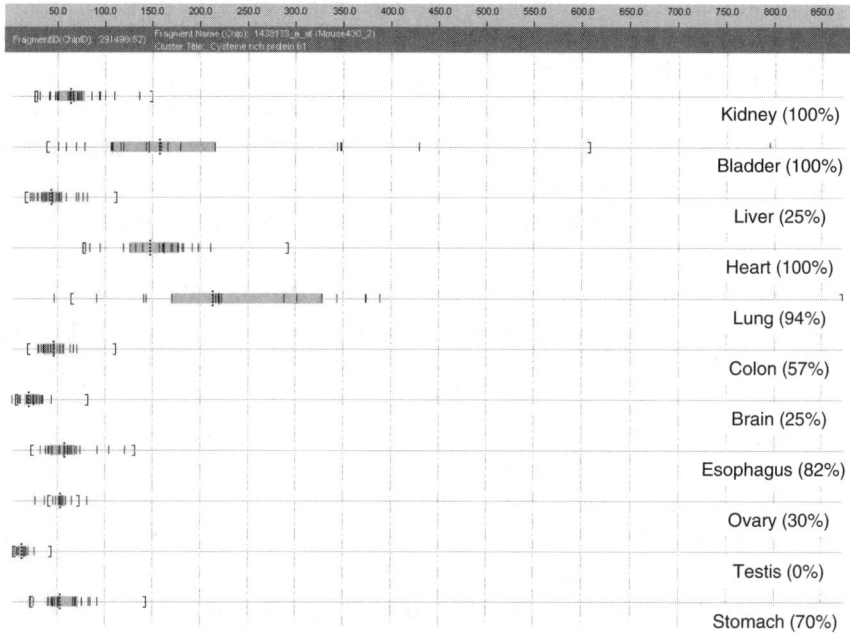

Fig. 18. e-Northern analysis of CYR61 in normal mouse tissues. CYR61 is represented by fragment 1438133_a_at on the Affymetrix GeneChip Mouse Genome 430 2.0 Array. Many tissues expressed CYR61 particularly the kidney, bladder, heart, and lung. For details on the e-Northern visual, please see the **Fig. 1** legend.

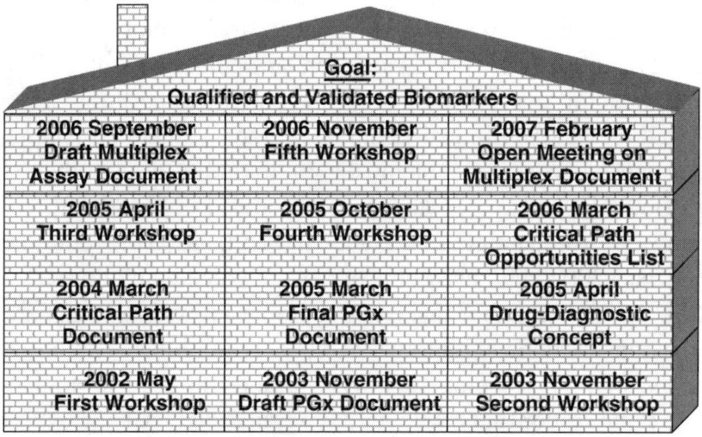

Goal:
Qualified and Validated Biomarkers
2006 September Draft Multiplex Assay Document
2006 November Fifth Workshop
2007 February Open Meeting on Multiplex Document
2005 April Third Workshop
2005 October Fourth Workshop
2006 March Critical Path Opportunities List
2004 March Critical Path Document
2005 March Final PGx Document
2005 April Drug-Diagnostic Concept
2002 May First Workshop
2003 November Draft PGx Document
2003 November Second Workshop

Fig. 19. This figure illustrates some of the building blocks the FDA has employed to encourage and enable the pharmaceutical industry to use pharmacogenomics in biomarker discovery and their qualification.

development for the advancement of safe and efficacious new medical products and to improve the efficiency of the process thereby reducing costs. A year later (March 2005), the finalized pharmacogenomic guidance document was released quickly followed by the Drug-Diagnostic Co-Development Concept Paper (http://www.fda.gov/cder/genomics/pharmacoconceptfn.pdf) in April 2005. By this time, it had become clear that if one develops biomarkers required to prescribe a drug (e.g., Herceptin® known generically as (trastuzumab), made by Genentech, South San Francisco, CA), then a testing kit needs to be developed at the same time as the drug. Concerns discussed at this third FDA-sponsored meeting in the same month focused on clinical applications including a need to clarify biomarker qualification (clinical utility) and analytical validation, the timing differences between using pharmacogenomics to identify biomarkers during drug development and approval of any final testing kit that may need to be implemented for drug approval, international regulatory differences, appropriate validation techniques used for biomarker qualification, and so forth. A series of papers came out of that meeting describing much of the discussion *(40–45)*. In October 2005, a fourth workshop was held to discuss the validation of biomarkers in safety and efficacy. Approximately 1 year later, a fifth workshop was convened and dealt primarily with genomic standards and submission of biomarker data to the FDA. It can be said that the promise of a genomics approach is no longer being questioned as much as the pathways used to identify and validate biomarkers. In March 2006, the FDA released their Critical Path Opportunities Report as a follow-up to their Critical Path Document, and here they list 76 areas of unmet need and profile biomarkers as the first (http://www.fda.gov/oc/initiatives/criticalpath/reports/opp_report.pdf). Some people believe the FDA took a step back from encouraging the development of pharmacogenomic biomarkers when, in September 2006, the Office of In Vitro Diagnostic Device Evaluation and Safety at the FDA released a draft guidance for oversight of some so-called home-brew tests that had been regulated by CLIA (Clinical Laboratory Improvement Amendments). The FDA has proposed to review diagnostic multivariate assays that employ an algorithm, something many genomic biomarker assays likely will require. This draft guidance was the subject of a somewhat contentious meeting held by the FDA in February 2007, and many concerns were voiced including the additional cost and time to market, a fear that existing home-brew assays that are assisting clinical decisions will be withdrawn if such requirements are finalized, the legality of the FDA overseeing such assays, and the need to amend the draft guidance to remove conflicts with CLIA regulations. All are waiting to see what new regulations are put into place to understand if and how they might stunt the growth of the new area of pharmacogenomic tests in the clinical arena.

As noted earlier, there is no widely accepted road that can be followed to ensure biomarker qualification, and the FDA's pharmacogenomic guidance document did not address this issue. To correct that situation, Goodsaid and Frueh outlined a proposed map of biomarker qualification and are working with consortia and pharmaceutical companies directly to determine if such a roadmap would be acceptable to the community at large *(46)*.

5. Conclusion

Genomic approaches have been used to identify potential translational biomarkers such as adipsin, a potential biomarker of gastrointestinal injury, and renal biomarkers including Kim-1, lipocalin-2, and CYR61. The advent of microarray technology that enables the assessment of tens of thousands of genes in a single sample and the generation of large reference databases of genomic information ensure that biomarker discovery will advance at faster speeds than ever before. Examples were provided of how large reference databases can assist in the discovery and qualification of biomarkers. However, this does not mean that the path to their qualification and validation is clear as scientists must agree on the approaches required to implement any biomarker for there to be a true increase in their usefulness. Genomic information can begin to shed light on the specificity of biomarkers' disease and toxicity and tissue localization thus narrowing the effort spent on even preliminary steps for qualification. The role of the FDA in advancing this area of research is critical to enable successful use of new biomarkers to meet safety and efficacy demands and improve the productivity of drug development.

References

1. Vaidya, V.S. and Bonventre, J.V. (2006) Mechanistic biomarkers for cytotoxic acute kidney injury. *Expert. Opin. Drug Metab. Toxicol.* **2**, 697–713.
2. Ichimura, T., Bonventre, J.V., Bailly, V., Wei, H., Hession, C.A., Cate, R.L., and Sanicola, M. (1998) Kidney injury molecule-1 (KIM-1), a putative epithelial cell adhesion molecule containing a novel immunoglobulin domain, is up-regulated in renal cells after injury. *J. Biol. Chem.* **273**, 4135–4142.
3. Amin, R.P., Vickers, A.E., Sistare, F., Thompson, K.L., Roman, R.J., Lawton, M., et al. (2004) Identification of putative gene based markers of renal toxicity. *Environ. Health Perspect.* **112**, 465–479.
4. Bailly, V., Zhang, Z., Meier, W., Cate, R., Sanicola, M., and Bonventre, J.V. (2002) Shedding of kidney injury molecule-1, a putative adhesion protein involved in renal regeneration. *J. Biol. Chem.* **277**, 39739–39748.

5. Davis, J.W., Goodsaid, F.M., Bral, C.M., Obert, L.A., Mandakas, G., Garner, C.E., et al. (2004) Quantitative gene expression analysis in a nonhuman primate model of antibiotic-induced nephrotoxicity. *Toxicol. Appl. Pharmacol.* **200**, 16–26.

6. Han, W.K., Bailly, V., Abichandani, R., Thadhani, R., and Bonventre, J.V. (2002) Kidney injury molecule-1 (KIM-1): a novel biomarker for human renal proximal tubule injury. *Kidney Int.* **62**, 237–244.

7. Ichimura, T., Hung, C.C., Yang, S.A., Stevens, J.L., and Bonventre, J.V. (2004) Kidney injury molecule-1: a tissue and urinary biomarker for nephrotoxicant-induced renal injury. *Am. J. Physiol. Renal Physiol.* **286**, F552–F563.

8. Kuehn, E.W., Park, K.M., Somlo, S., and Bonventre, J.V. (2002) Kidney injury molecule-1 expression in murine polycystic kidney disease. *Am. J. Physiol. Renal Physiol.* **283**, F1326–F1336.

9. Liangos, O., Perianayagam, M.C., Vaidya, V.S., Han, W.K., Wald, R., Tighiouart, H., et al. (2007) Urinary N-acetyl-beta-(D)-Glucosaminidase activity and kidney injury molecule-1 level are associated with adverse outcomes in acute renal failure. *J. Am. Soc. Nephrol.* **18**, 904–912.

10. Vaidya, V.S., Ramirez, V., Ichimura, T., Bobadilla, N.A., and Bonventre, J.V. (2006) Urinary kidney injury molecule-1: a sensitive quantitative biomarker for early detection of kidney tubular injury. *Am. J. Physiol. Renal Physiol.* **290**, F517–F529.

11. Olson, H., Betton, G., Robinson, D., Thomas, K., Monro, A., Kolaja, G., et al. (2000) Concordance of the toxicity of pharmaceuticals in humans and in animals. *Regul. Toxicol. Pharmacol.* **32**, 56–67.

12. Evans, R.W., Williams, G.E., Baron, H.M., Deng, M.C., Eisen, H.J., Hunt, S.A., et al. (2005) The economic implications of noninvasive molecular testing for cardiac allograft rejection. *Am. J. Transplant.* **5**, 1553–1558.

13. Starling, R.C., Pham, M., Valantine, H., Miller, L., Eisen, H., Rodriguez, E.R., et al. (2006) Molecular testing in the management of cardiac transplant recipients: initial clinical experience. *J. Heart Lung Transplant.* **25**, 1389–1395.

14. Baron, C., Somogyi, R., Greller, L.D., Rineau, V., Wilkinson, P., Cho, C.R., et al. (2007) Prediction of graft-versus-host disease in humans by donor gene-expression profiling. *PLoS Med.* **4**, e23.

15. Baird, A.E. (2007) Blood genomics in human stroke. *Stroke* **38**, 694–698.

16. Burczynski, M.E. and Dorner, A.J. (2006) Transcriptional profiling of peripheral blood cells in clinical pharmacogenomic studies. *Pharmacogenomics* **7**, 187–202.

17. McHale, C.M., Zhang, L., Hubbard, A.E., Zhao, X., Baccarelli, A., Pesatori, A.C., et al. (2007) Microarray analysis of gene expression in peripheral blood mononuclear cells from dioxin-exposed human subjects. *Toxicology* **229**, 101–113.

18. Bushel, P.R., Heinloth, A.N., Li, J., Huang, L., Chou, J.W., Boorman, G.A. et al. (2007) Blood gene expression signatures predict exposure levels. *Proc. Natl. Acad. Sci U.S.A* **104**, 18211–18216.

19. Searfoss, G.H., Jordan, W.H., Calligaro, D.O., Galbreath, E.J., Schirtzinger, L.M., Berridge, B.R., et al. (2003) Adipsin, a biomarker of gastrointestinal toxicity mediated by a functional gamma-secretase inhibitor. *J. Biol. Chem.* **278**, 46107–46116.

20. Milano, J., McKay, J., Dagenais, C., Foster-Brown, L., Pognan, F., Gadient, R., et al. (2004) Modulation of notch processing by gamma-secretase inhibitors causes intestinal goblet cell metaplasia and induction of genes known to specify gut secretory lineage differentiation. *Toxicol. Sci.* **82**, 341–358.
21. Lin, F., Zhang, P.L., Yang, X.J., Shi, J., Blasick, T., Han, W.K., et al. (2007) Human kidney injury molecule-1 (hKIM-1): a useful immunohistochemical marker for diagnosing renal cell carcinoma and ovarian clear cell carcinoma. *Am. J. Surg. Pathol.* **31**, 371–381.
22. Mendrick, D.L. and Daniels, K.K. (2007) From the bench to the clinic and back again: translational biomarker discovery using *in silico* mining of pharmacogenomic data. *Biomarkers Med.* **1**, 319–333.
23. Gielen, A.W., Lobell, A., Lidman, O., Khademi, M., Olsson, T., and Piehl, F. (2005) Expression of T cell immunoglobulin- and mucin-domain-containing molecules-1 and -3 (TIM-1 and -3) in the rat nervous and immune systems. *J. Neuroimmunol.* **164**, 93–104.
24. de Souza, A.J., Oriss, T.B., O'malley, K.J., Ray, A., and Kane, L.P. (2005) T cell Ig and mucin 1 (TIM-1) is expressed on in vivo-activated T cells and provides a costimulatory signal for T cell activation. *Proc. Natl. Acad. Sci U. S. A.* **102**, 17113–17118.
25. Meyers, J.H., Sabatos, C.A., Chakravarti, S., and Kuchroo, V.K. (2005) The TIM gene family regulates autoimmune and allergic diseases. *Trends Mol. Med.* **11**, 362–369.
26. Thompson, K.L., Afshari, C.A., Amin, R.P., Bertram, T.A., Car, B., Cunningham, M., et al. (2004) Identification of platform-independent gene expression markers of cisplatin nephrotoxicity. *Environ. Health Perspect.* **112**, 488–494.
27. Emanuelli, G., Anfossi, G., Calcamuggi, G., Marcarino, C., Ottone, G., and Dughera, L. (1985) Urinary enzyme excretion in acute and subacute experimental lithium administration. *Enzyme* **34**, 177–185.
28. Cornin, R.E. and Henrich, W.L.(1996) Toxic nephropathy. In: *The Kidney* (Brenner, B.M., ed.), W.B. Saunders, Philadelphia, pp. 1680–1711.
29. Goldstein, R.S. and Schnellmann, R.G.(1996) Toxic responses of the kidney. In: *Toxicology. The Basic Science of Poisons* (Klaassen, C.D., ed.), McGraw-Hill, New York, pp. 417–442.
30. Friedl, A., Stoesz, S.P., Buckley, P., and Gould, M.N. (1999) Neutrophil gelatinase-associated lipocalin in normal and neoplastic human tissues. Cell type-specific pattern of expression. *Histochem. J.* **31**, 433–441.
31. Mishra, J., Ma, Q., Prada, A., Mitsnefes, M., Zahedi, K., Yang, J., et al. (2003) Identification of neutrophil gelatinase-associated lipocalin as a novel early urinary biomarker for ischemic renal injury. *J. Am. Soc. Nephrol.* **14**, 2534–2543.
32. Schmidt-Ott, K.M., Mori, K., Li, J.Y., Kalandadze, A., Cohen, D.J., Devarajan, P., and Barasch, J. (2007) Dual action of neutrophil gelatinase-associated lipocalin. *J. Am. Soc. Nephrol.* **18**, 407–413.

33. Lau, L.F. and Lam, S.C. (1999) The CCN family of angiogenic regulators: the integrin connection. *Exp. Cell Res.* **248**, 44–57.
34. Menendez, J.A., Mehmi, I., Griggs, D.W., and Lupu, R. (2003) The angiogenic factor CYR61 in breast cancer: molecular pathology and therapeutic perspectives. *Endocr. Relat. Cancer* **10**, 141–152.
35. Muramatsu, Y., Tsujie, M., Kohda, Y., Pham, B., Perantoni, A.O., Zhao, H., et al. (2002) Early detection of cysteine rich protein 61 (CYR61, CCN1) in urine following renal ischemic reperfusion injury. *Kidney Int.* **62**, 1601–1610.
36. Salerno, R.A. and Lesko, L.J. (2004) Pharmacogenomics in drug development and regulatory decision-making: the Genomic Data Submission (GDS) Proposal. *Pharmacogenomics* **5**, 25–30.
37. Leighton, J.K., DeGeorge, J., Jacobson-Kram, D., MacGregor, J., Mendrick, D., and Worobec, A. (2004) Pharmacogenomic data submissions to the FDA: non-clinical case studies. *Pharmacogenomics* **5**, 507–511.
38. Trepicchio, W.L., Williams, G.A., Essayan, D., Hall, S.T., Harty, L.C., Shaw, P.M., et al. (2004) Pharmacogenomic data submissions to the FDA: clinical case studies. *Pharmacogenomics* **5**, 519–524.
39. Ruano, G., Collins, J.M., Dorner, A.J., Wang, S.J., Guerciolini, R., and Huang, S.M. (2004) Pharmacogenomic data submissions to the FDA: clinical pharmacology case studies. *Pharmacogenomics* **5**, 513–517.
40. Mendrick, D.L., Brazell, C., Mansfield, E.A., Pietrusko, R., Barilero, I., Hackett, J., et al. (2006) Pharmacogenomics and regulatory decision making: an international perspective. *Pharmacogenomics J.* **6**, 154–157.
41. Salerno, R.A. and Lesko, L.J. (2006) Three years of promise, proposals, and progress on optimizing the benefit/risk of medicines: a commentary on the 3rd FDA-DIA-PWG-PhRMA-BIO pharmacogenomics workshop. *Pharmacogenomics J.* **6**, 78–81.
42. Frueh, F.W., Rudman, A., Simon, K., Gutman, S., Reed, C., and Dorner, A.J. (2006) Experience with voluntary and required genomic data submissions to the FDA: summary report from track 1 of the third FDA-DIA-PWG-PhRMA-BIO pharmacogenomics workshop. *Pharmacogenomics J.* **6**, 296–300.
43. Simon, R. and Wang, S.J. (2006) Use of genomic signatures in therapeutics development in oncology and other diseases. *Pharmacogenomics J.* **6**, 166–173.
44. Hinman, L.M., Huang, S.M., Hackett, J., Koch, W.H., Love, P.Y., Pennello, G., et al. (2006) The drug diagnostic co-development concept paper Commentary from the 3rd FDA-DIA-PWG-PhRMA-BIO Pharmacogenomics Workshop. *Pharmacogenomics J.* **6**, 375–380.
45. Trepicchio, W.L., Essayan, D., Hall, S.T., Schechter, G., Tezak, Z., Wang, S.J., et al. (2006) Designing prospective clinical pharmacogenomic (PG) trials: meeting report on drug development strategies to enhance therapeutic decision making. *Pharmacogenomics J.* **6**, 89–94.
46. Goodsaid, F. and Frueh, F. (2006) Process map proposal for the validation of genomic biomarkers. *Pharmacogenomics* **7**, 773–782.

11

Public Consortium Efforts in Toxicogenomics

William B. Mattes

Summary

Public consortia provide a forum for addressing questions requiring more resources than one organization alone could bring to bear and engaging many sectors of the scientific community. They are particular well suited for tackling some of the questions encountered in the field of toxicogenomics, where the number of studies and microarray analyses would be prohibitively expensive for a single organization to carry out. Five consortia that stand out in the field of toxicogenomics are the Institutional Life Sciences Institute (ILSI) Health and Environmental Sciences Institute (HESI) Committee on the Application of Genomics to Mechanism Based Risk Assessment, the Toxicogenomics Research Consortium, the MicroArray Quality Control (MAQC) Consortium, the InnoMed PredTox effort, and the Predictive Safety Testing Consortium. Collectively, these consortia efforts have addressed issues such as reproducibility of microarray results, standard practice for assays and analysis, relevance of microarray results to conventional end points, and robustness of statistical models on diverse data sets. Their results demonstrate the impact that the pooling of resources, experience, expertise, and insight found in consortia can have.

Key Words: consortia; public-private partnerships; standards; toxicogenomics.

1. Introduction

With the description of the first microarray in 1995 *(1,2)*, the field of transcriptional profiling was born, where hundreds to thousands of gene expression changes are monitored simultaneously. This advance was not the first approach to examining changes in gene expression on a large scale: two-dimensional gel electrophoresis monitored global changes in levels of discrete proteins *(3)*, and differential display had the capability of detecting changes in unique transcripts *(4)*. However, both of these techniques monitor changes

From: *Methods in Molecular Biology, vol. 460: Essential Concepts in Toxicogenomics*
Edited by: D. L. Mendrick and W. B. Mattes © Humana Press, Totowa, NJ

in unknown proteins or transcripts on electrophoretograms; the identity of the proteins or mRNAs must be determined by subsequent analysis. In contrast, the identity of each element on a microarray is known such that signals are immediately associated with a known transcript. The potential for applying such analyses to understanding responses to toxicants was rapidly appreciated *(5–10)*. On the other hand, the variety of mircroarray formats *(11–13)* and questions as to appropriate protocols and analyses *(14–17)*, the reproducibility of the data produced, and the overall applicability to standard toxicology studies generated a desire for coordinated, public efforts to determine the true value and standard practice of transcript profiling as applied to toxicology. Indeed, over the past 8 years or so, several different collaborative efforts have communally explored questions relevant to the practice of toxicogenomics. Ultimately, these efforts led to more universal understanding of the true pitfalls and potentials of this field and its effective application to the science of toxicology.

2. Consortia and other Public Efforts

2.1. ILSI HESI Committee on the Application of Genomics to Mechanism Based Risk Assessment

2.1.1. Background and Goals

The ILSI is a nonprofit foundation whose "goal is to further the understanding of scientific issues relating to nutrition, food safety, toxicology, risk assessment, and the environment by bringing together scientists from academia, government, and industry" *(18)*. The ILSI HESI is the global branch of ILSI, and in mid-1999 its membership "formed the Committee on the Application of Genomics to Mechanism Based Risk Assessment to develop a collaborative scientific program to address issues, challenges, and opportunities afforded by the emerging field of toxicogenomics" *(19)*. In particular, the committee initially focused on the issue of concordance of results collected on different platforms and by different investigators (i.e., the sources of variability). Later projects, begun after 2004, have focused on mechanism-based markers of toxicity, a public baseline gene expression database, a determination of standard practice among companies using toxicogenomics, and development of gene expression profile capable of differentiating cytotoxic, DNA reactive, and DNA nonreactive genotoxic mechanisms for compounds with positive findings in the *in vitro* chromosome damage assays.

2.1.2. Organization and Approaches

The committee ultimately was made up of 30 corporate participants including representatives from the pharmaceutical, agrochemical, chemical, and consumer products sectors, as well as scientists from governmental and academic laboratories. From its earliest inception, the committee saw the value of addressing different areas of toxicity and experimental design, as well as the need to develop a repository or database for storing the raw data generated by the various experiments. Thus four working groups were set up to address nephrotoxicity, hepatotoxicity, genotoxicity, and the issue of database development; the first three listed working groups developed experimental protocols that would investigate sources of biological and technical variability, resulting in the matrix of experiments outlined in **Table 1**. The Database Working Group developed specifications for a toxicogenomics database that could store genomic data from different platforms as well as the commonly collected in-life data such as clinical pathology and histopathology. Theses specifications were used to identify the European Bioinformatics Institute (EBI) as a partner in database development (see Ref. *(20)*). Even so, an interim solution consisting of a simple format for data sharing, the ILSI Microarray Database (IMD) *(21)*, was established.. The IMD used a "Basic Expression Exchange Format," developed by Detlef Wolf of Roche, which called for microarray data to be submitted as text files with only probe identifiers and fold-change values. Other information, unique to a platform or experiment, could also be submitted as part of the file. Sequence information for the probes was provided as a separate file, allowing annotation updates. Thus the IMD allowed for simple data comparisons while the more comprehensive database was developed by EBI. .

2.1.3. Outcomes

The work of the committee was reported in a number of articles as part of a Mini-Monograph in Environmental Health Perspectives, and an overview

Table 1
ILSI HESI Committee on the Application of Genomics

	Working group		
	Nephrotoxicity	Hepatotoxicity	Genotoxicity
Compounds	3	2	11
Study sites	1/compound	2/compound	
Doses (including vehicle)	4	3	
Time points	4	5	2
Experimental system	Male rats	Male rats	2 cell lines
Array platforms	5	3	10
Array sites	10	20	16

was provided in the lead article by Pennie, Pettit, and Lord *(22)*. The results serve as a landmark example of how difficult questions can be addressed by consortium efforts, notably the following:

2.1.3.1. BIOLOGICAL VARIABILITY (INTERSTUDY AND INTER-INDIVIDUAL)

The experimental results of the committee affirmed the historical truth that individual animals can respond differently to same treatments and that the same study design may elicit quantitatively different results in different laboratories. Thus whereas the two studies with the same protocol for treatment with clofibrate both produced the expected signs of hepatotoxicity, the high-dose treatment group in one study exhibited on average a twofold greater response in terms of liver weight changes, clinical chemistry, and liver acyl-CoA oxidase activity *(23,24)*. Similarly, animals treated in a pilot study examining cisplatin-induced nephrotoxicity elicited a much less pronounced response on the average than those treated with a similar regime (5 mg/kg for 144 h) in a subsequent study *(25,26)*. Interestingly enough, for a limited subset of transcripts analyzed, a similar trend in gene expression change was observed for both replicate experiments even though the serum markers of nephrotoxicity were altered to different levels *(25)*, and gene expression data for the two replicates, as measured by a cDNA microarray, clustered together *(27)*. Such observations would suggest that the gene expression responses were more robust than traditional biological parameters. On the other hand, individual animals responded quite differently to the 5 mg/kg, 144-h cisplatin treatment, with one of the five showing no response in at all in conventional clinical chemistry and histopathology *(25)*. These individual responses measured with conventional parameters were mirrored at the gene expression level, and the animal showing no histopathologic response also showed no gene expression response *(25)*.

2.1.3.2. POOLING OF RNA SAMPLES VERSUS MEASURING GENE EXPRESSION INDIVIDUAL BY INDIVIDUAL

Whereas a large majority of the microarray experiments conducted by the committee used RNA pooled from all the animals in a given treatment group, the experience noted above highlighted a conclusion reached toward the end of the committee's work: "Running samples from multiple individual animals is more costly than using pooled samples. However, it is advantageous in that it can be used to identify outliers and also permits the calculation of statistically significant changes in gene expression …. Such experiments, although more time and resource intensive than the use of pooled samples, permit analysis of interindividual (biological) variation" *(23)*. Note that outliers can be either hyperresponders or nonresponders, and their identification can lead to a more informed analysis. Furthermore, the statistical approach to analyses of gene expression among individual animals in an experimental group then can mirror the statistical approach to analysis of conventional parameters in that group. Indeed, these themes have been further explored and affirmed in later publications *(28)*.

2.1.3.3. TECHNICAL VARIABILITY (INTERLABORATORY)

Several of the studies conducted by the committee examined the effects of analyzing the same samples in different laboratories but using a single microarray platform. In the case of cDNA arrays "…despite some quantitative differences …there was generally good concordance between" the data obtained in different laboratories *(23)*. Technical differences between array handling and processing were speculated to be the source of the quantitative differences. Likewise, whereas considerable interlaboratory variability was noted in the results with Affymetrix microarrays, this variability did not obscure the clear difference in gene expression between control and treated samples, nor did it "impair the detection of molecular effects elicited" by treatment *(29)*. Indeed a rigorous statistical analysis demonstrated that site effects could be adjusted such that the results from any given site told the same story in terms of the significantly differentially expressed genes *(30)*.

2.1.3.4. TECHNICAL VARIABILITY (INTERPLATFORM)

Obviously, a key question addressed in the committee's efforts was to what extent the biological story indicated by gene expressions changes identified with one microarray platform was congruent with that identified with another platform. The encouraging outcome was that indeed certain major genes and pathways are observed as regulated by toxicant treatment regardless of platform. Thus with methapyrilene treatment, upregulation of liver transcripts for glutathione-*S*-transferase (Yc2), aflatoxin B1-reductase (Akr7a3), and heme oxygenase 1 (Hmox1) with downregulation of liver transcripts for fatty acid metabolism enzymes and several cytochrome P450s were seen regardless whether a cDNA *(31)* or Affymetrix *(29)* platform was used. Similarly, 43 kidney transcripts could be identified as regulated after cisplatin treatment regardless of whether three different cDNA microarray platforms or an Affymetrix microarray platform was used *(25)*.

2.1.3.5. MICROARRAY AND GENE ANNOTATION

Early on in the committee's work, it became clear that a significant stumbling block both to cross-platform comparisons and to overall biological interpretation was the uniform identification and annotation of the transcripts (i.e., genes) queried by any given microarray. Biologists will often use nonstandard gene names and symbols, and these can find their way into microarray annotations, particularly in those used at the time the committee was conducting experiments. Thus an informatic approach to standardized annotation was developed *(32)*. On the other hand, certain apparent discrepancies between results on different microarray platforms were found to be due to subtle but important differences between the sequences queried by the different platforms. Thus differences in the sequences annotated in two different platforms as caldesmon led to dramatically different gene regulation results and highlighted the need to address questionable results with quantitative PCR *(33)*.

2.2. The Toxicogenomics Research Consortium

2.2.1. Background and Goals

The Toxicogenomics Research Consortium (TRC) arose out of a Request for Applications issued by the National Institutes of Environmental Health in November 2000 and was established in November 2001. It envisioned that selected participants would work on coordinated, multidisciplinary toxicogenomics research with the support of NIEHS extramural staff, the NIEHS National Center for Toxicogenomics (NCT), and NCT-supported Resource Contractors. Program goals included:

1. Enhancement of research in the broad area of environmental stress responses using microarray gene expression profiling;
2. Development of standards and practices which will allow analysis of gene expression data across platforms and provide an understanding of intra- and inter-laboratory variation;
3. Contribution to the development of a robust relational database which combines toxicological endpoints with changes in gene expression profiles;
4. Improvement of public health through better risk detection, and earlier intervention in disease processes" *(34)*.

As with other consortia, the hope was that synchronization of research efforts of many scientists in various locations promises to produce high-quality results faster and more reliably than isolated research teams working alone *(35)*.

2.2.2. Organization and Approaches

The Toxicogenomics Research Consortium consisted of six Cooperative Research Members (CRMs), with five academic centers (**Table 2**) as well as the NIEHS Microarray Group. The consortium activities were also supported by two NIEHS NCT Resource Contractors: Icoria (now CoGenics) providing microarray support and Science Applications International Corporation (SAIC) providing bioinformatics support. In addition to investigator-initiated research projects, each academic CRM had a Toxicology Research Core Project

Table 2
Organization of the TRC

Academic CRM	Toxicology Research Core project
Duke University	Transition metal response genes in *C. elegans.*
Fred Hutchinson Cancer Research Center/ University of Washington	Glutamate-cysteine ligase overexpression and the potential attenuation of carbon tetrachloride and methyl mercury induced oxidative stress in liver and kidney.
Massachusetts Institute of Technology	Interindividual differences in the transcriptional responsiveness of 24 human lymphoblastoid cell lines after exposure to environmental agents.
Oregon Health and Science University	Gene expression analysis after exposure to neurotoxic aromatics (neurotoxic 1,2-DAB vs. non-neurotoxic 1,3-DAB) and aliphatics (neurotoxic 2,5-HD vs. non-neurotoxic 2,4-HD).
University of North Carolina at Chapel Hill	Nuclear receptor-mediated toxicity and gene expression profiling *in vitro* and *in vivo* with mice lacking AhR, PPAR-a, and CAR.

(**Table 2**) through which the dependent, collaborative toxicology work was performed under the Cooperative Research Program. Thus, each center had the dual function of performing cutting-edge independent research employing microarray gene expression profiling, while also performing collaborative toxicology experiments for interlaboratory standardization and validation.

To systematically address the questions of interlaboratory and interplatform reproducibility in microarray data, the TRC took the approach of focusing on two different RNA samples, a mouse liver RNA sample and a pooled RNA sample, prepared from equal amounts of mouse liver, kidney, lung, brain, and spleen RNA. Aliquots of these two RNA samples were distributed to seven laboratories, with analyses performed on a total of 12 microarray platforms. All seven laboratories used two of the microarray platforms as "standard" platforms, these being a spotted long oligonucleotide microarray produced by one consortium member and a commercially produced long oligonucleotide microarray.

2.2.3. Outcomes

A striking, albeit predictable, conclusion from the validation experiments was that if experiments using the same microarray platform are to be compared between laboratories at the level of signal intensity, RNA labeling, hybridization, file format, image analysis, and nomenclature must be completely standardized.

All of these factors contribute to interlaboratory variability, although standardization of RNA labeling and hybridization appeared to offer the greatest reduction in variability. Similar comparisons (using the same microarray platform), but at the level of expression ratios, demanded standardization of the same parameters, and included standardization of the normalization method (Lowess without background subtraction). As with signal intensity, standardization of RNA labeling and hybridization had the greatest effect in reducing variability.

Of course, an important question is how experiments using different microarray platforms may be compared with each other. First and foremost is the question of how comparable the sequences queried by different microarrays are. For the 12 platforms used in the TRC experiments, only 502 transcripts could be unambiguously identified as being in common. The expression ratios for these transcripts, determined with the same commercial platform, were reasonably comparable within and across different laboratories. However, comparisons between microarray platforms were quite poor, and especially so when compared between laboratories. On the other hand, when expression data was analyzed using the EASE algorithm *(36)* for Gene Ontology *(37)* groups regulated, many similar themes were observed regardless of microarray platform or laboratory. Overall, the results of the Toxicogenomics Research Consortium emphasized the importance of standardization of experimental factors for comparisons of results.

2.3. MicroArray Quality Control (MAQC) Consortium

2.3.1. Background, Goals, and Organization

The MicroArray Quality Control Consortium was launched in February 2005, in order to address reliability concerns as well as other performance, standards, quality, and data analysis issues. It envisioned its work as providing a framework for regulatory and clinical use of microarray data. Under direction of scientists at the FDA's National Center for Toxicological Research (NCTR), the consortium brought together six FDA centers, major providers of microarray platforms and RNA samples, the Environmental Protection Agency (EPA), the National Institute of Standards and Technology (NIST), several academic laboratories, and other stakeholders (**Table 3**). To achieve its goals, the consortium developed four pools composed of two RNA sample types to use for assessing platform performance, technical processing, and data analysis *(38)* and developed a data set with more than 1300 microarrays from more than 40 test sites and 20 microarray platforms. Furthermore, comparative performance assessments were made possible by mapping microarray proves to RefSeq and AceView databases and identifying for each platform 12,091 probes matching a common set of 12,091 reference sequences from 12,091 different genes *(39)*.

Table 3
Participants in the MAQC Consortium

Academic centers	Public sector institutions	Private sector organizations
Burnham Institute	EPA	Affymetrix
Cold Spring Harbor Laboratory	FDA/CBER	Agilent
Duke University	FDA/CDER	Ambion
Harvard/Children's Hospital	FDA/CDRH	Applied Biosystems
MD Anderson	FDA/CFSAN	Biogen Idec
Stanford University	FDA/CVM	CapialBio
UCLA/Cedars-Sinai	FDA/NCTR	Clontech
UCSF	NIH/NCBI	Combimatrix
UIUC	NIH/NCI	Eppendorf
UMass Boston	NIST	Expression Analysis
Vanderbilt University		Full Moon BioSystems
Wake Forest University		GE Healthcare
Yale University		Gene Express, Inc.
		Genospectra
		GenUs BioSystems
		Icoria
		Illumina
		Novartis
		SAS
		Stratagene
		TeleChem ArrayIt
		ViaLogy

Major findings of the first phase of the MAQC project were published in six research papers in the September 8, 2006, issue of *Nature Biotechnology*.

2.3.2. Outcomes

One of the efforts was to establish confidence in the quantitative nature of microarray results. Thus three quantitative gene expression technologies were used to test the results of five microarray platforms, with the conclusion that the numerical gene expression values determined by all platforms were comparable, except where there were differences in probe sequence and thus target location. Weakly expressed genes also detracted from comparability of results *(40)*. Furthermore, a rigorous comparison of one-color and two-color microarray platforms demonstrated high concordance of results. As an example, the lists of differentially expressed gene lists are consistent across one- and two-color platforms when using widely accepted *p*-value and fold-change thresholds for significance *(41)*. A critical finding coming out of this effort centered on analysis of microarray results. The MAQC work demonstrated that the apparent lack of reproducibility reported in previous studies using microarray assays was likely caused, at least in part, by the common practice of ranking genes solely by a statistical significance measure, for example, *p* values derived from simple *t*-tests, and selecting differentially expressed genes with a stringent significance threshold. Fold change ranking plus a nonstringent *p*-value cutoff can be used as a baseline practice for generating more reproducible signature gene lists *(39,42)*. This approach also minimized the effects of different methods of normalization, or global scaling methods, which do affect gene lists ranked solely by statistical significance. The MAQC results suggest that microarray data analysis for the identification of reproducible lists of differentially expressed genes does not need be complicated or confusing and can lend itself to consensus. Thus with robust statistical analyses and approaches for assessing technical performance and proficiency of the microarray assays, the MAQC project provided a foundation for the routine use of these tools in a regulated setting such as clinical diagnostics.

2.4. InnoMed PredTox

2.4.1. Background, Organization, and Goals

InnoMed PredTox, founded in 2005, is a joint Industry and European Commission research consortium funded by the European Union under an advisory agreement with European Federation of Pharmaceutical Industries and Associations (EFPIA) with a goal of improving the prediction of drug safety *(43)*. The consortium is composed of 14 pharmaceutical companies, 3 academic institutions, and firms providing computational and laboratory resources (**Table 4**). The questions the project hope to answer include:

- "What combination of methods/technologies delivers the best predictive results for hepatotoxicity and / or nephrotoxicity?"
- "What combination of methods/technologies delivers the best predictive results for each individual compound?"
- "What combination of methods delivers the best predictive results overall?"
- "Have 'new' candidate biomarkers of hepatotoxicity and/or nephrotoxicity been identified and validated?"
- "Can an ideal mix of methods/technologies be recommended for future drug development projects?"
- "What is the predictive value for extrapolation to longer term rat studies?"
- "What is the predictive value for extrapolation to humans?"
- "What is the added value compared to conventional methods and approaches?"

The approach the consortium has taken has been to focus on a few proprietary compounds as examples and conduct studies where a variety of data types are

Table 4
Participants in the PredTox Consortium

Private sector	Academic
Bayer Schering Pharma	UCD Conway Institute
Bio-Rad	University of Hacettepe
Boehringer Ingelheim	Würzburg Institute of Toxicology
F. Hoffmann-La Roche	
Genedata	
Johnson & Johnson Pharmaceutical R & D	
Lilly S.A.	
Merck KGaA	
MerckSerono	
Novartis	
Novo Nordisk A/S, DK	
Nycomed	
Organon	
Sanofi-Aventis (Germany, France)	
Servier	

collected, including transcriptomics, proteomics, metabonomics, and conventional end points. The questions will then be explored with data mining of the integrated results. The PredTox work is organized into specific work packages (WPs) covering different aspects of the studies, each package being managed by a given work group. Thus compound selection, study design, gene expression, and so forth, are carried out by different work groups.

2.4.2. Outcomes

At this time, all in-life studies are complete, as well as transcriptomics and SELDI (Surface-enhanced laser desorption ionization) proteomics data collection. 2D-DIGE (2 Dimensional Difference Gel Electrophoresis) and 2D-PAGE (2 Dimensional Polyacrylamide Gel Electrophoresis) data collection is still in progress, as is that from LC/MS and NMR metabonomics. A database designed to house and query such integrated data has been developed as well. Reports of the progress of the consortium have been communicated in several conferences and can be viewed at the consortium's Web site *(43)*.

2.5. The Predictive Safety Testing Consortium

2.5.1. Background and Goals

The Predictive Safety Testing Consortium (PSTC) was established to work in collaboration with The Critical Path Institute ("C-Path") and the FDA to identify and validate safety biomarkers. Discussions between potential members began in mid-2005, shortly after C-Path (a nonprofit, publicly funded institute) was established. C-Path's mission is to facilitate collaborations among government regulators, the academic community, and regulated businesses to improve the process of drug development, and a particular focus are those areas of need mentioned in the FDA's Critical Path Opportunities list (44). The PSTC was envisioned to address two main goals: the need to establish criteria for biomarker qualification (Opportunity No. 1) and the need to "modernize predictive toxicology" (Opportunity No. 20). In the case of the former, the PSTC would work with the FDA to examine what data sets and documents serve to support formal qualification of biomarkers in a given context. The context for the latter effort is that predictive safety tests internally developed and used by each individual company are of limited value to the FDA because the methods used have not been validated by an independent party. Furthermore, individual companies often lack the resources to thoroughly develop and critically examine new safety assays. The consortium provides a mechanism for member companies to share preliminary data on such assays and communally approach more rigorous validation. Thus its stated goals include (1) to cross-qualify preclinical animal model biomarkers aimed at reducing the cost and time of preclinical safety studies, (2) to use the combined resources, sample sets, and novel compounds to generate a biomarker data package convincing enough for FDA qualification as an approved biomarker, (3) to provide potential early indicators of clinical safety in drug development and postmarketing surveillance (i.e., a long-term goal of translational biomarkers), and (4) to develop new tools for FDA to assist in regulatory decision making.

2.5.2. Organization and Approaches

The formation of the PSTC was announced on March 16, 2006 (45), and included 8 pharmaceutical company members. Since that time, the consortium has grown to include 16 pharmaceutical company members, as well as SRI International as a founding partner (**Table 5**). Representatives from both the FDA and the European Medicines Evaluation Agency (EMEA) serve as advisors to the consortium. The PSTC is currently organized into five working groups addressing different pathology areas, including nephrotoxicity, hepatotoxicity, carcinogenicity, myopathy, and vascular injury. Key to the PSTC's structure is a Consortium Agreement that as a legal document addresses key concerns such as membership, antitrust issues, governance, funding, information sharing, confidentiality, publicity, and intellectual property. Membership is open to organizations with significant expertise and programs in safety biomarkers (SBMs) and willing to (1) share data, know-how, intellectual

Table 5
Participants in the Predictive Safety Testing Consortium

Private sector	Public sector
Abbott	The Critical Path Institute
Amgen, Inc	SRI International
Astra Zeneca	FDA
Boehringer Ingelheim	EMEA (European Medicines Evaluation Agency)
Bristol-Myers Squibb	ClinXus
Eli Lilly, Inc	
GlaxoSmithKline	
Iconix Pharmaceuticals	
Johnson & Johnson Pharmaceutical R&D	
Merck & Co., Inc.	
Novartis	
Pfizer, Inc.	
Roche	
Sanofi-Aventis U.S. Inc	
Schering Plough Research Institute	
Wyeth	

property (IP) for goal of qualification, (2) commit internal resources for qualifying other member's SBMs, (3) make available IP to other members of the consortium, and (4) enter into the Consortium Agreement and Project Agreements. Project Agreements represent the specific legal documents covering working group research projects. Governance of the consortium is handled by an "Advisory Committee" where each member company has one vote. The Critical Path Institute serves as the management for the consortium. Aside from the five working groups there is a Data Management Subcommittee overseeing the data repository and informatic needs of the consortium and a Translational Team developing strategies for studies supporting the use of PSTC-qualified biomarkers in a clinical setting.

Although the legal title of the PSTC's agreement is the "Toxicogenomic Cross-Validation Consortium Agreement," most of the focus of the working groups has been on nongenomic, accessible protein and metabolite biomarkers. On the other hand, many of the working groups have sought to share and use toxicogenomics data to provide support for further work with a given biomarker. The Carcinogenicity Working Group serves as one case where genomic signatures of nongenotoxic hepatocarcinogenicity are explicitly being validated using member data. To the end of supporting biomarker data validation, a consortium data repository, based on the Rosetta Resolver software system, is being developed.

2.5.3. Outcomes

As noted above, a large focus of the consortium has been on nongenomic biomarkers, and to that end the consortium has made considerable progress. The Nephrotoxicity Working Group has examined a panel of 23 urinary protein assays and submitted qualification results for seven of these to the FDA and EMEA. The Hepatotoxicity Working Group is pursuing a number of assays, but has considerable progress focused on four serum enzyme assays. Even so, the Carcinogenicity Working Group has critically examined certain published genomic signatures of nongenotoxic carcinogenicity *(46,47)* by evaluating their performance with member company genomic data. The results of this examination have been encouraging enough to suggest that the signatures be reassessed using a common gene expression platform (e.g., quantitative RT-PCR).

3. Discussion and Perspectives

Clearly, the strength of public consortia is the ability to tackle questions requiring more resources than one organization alone could bring to bear and to do so in a forum that engages many sectors of the scientific community. Some of the questions encountered early in the development of toxicogenomics—reproducibility of microarray results, standard practice for assays and analysis, relevance to conventional end points, and so forth—required precisely the concentration of resources provided by consortia. It is hard to imagine how the number of studies and microarray experiments required to rigorously address these questions could have been mustered by one organization. Importantly, consortia also provide a concentration and diversity of experience, expertise, and insight that are necessary to design and interpret the studies required to answer these questions.

The consortia efforts described all provided forums for moving the field of toxicogenomics forward. Thus we now have clear information on the strengths and limitations of various microarray platforms. At one time, it was hypothesized that microarray measurements would be dramatically more sensitive than conventional end points (i.e., detect meaningful changes in physiologic states under treatment conditions where conventional end points never showed any responses) and that such measurements would be too noisy to be useful (i.e., susceptible to biological fluctuation). The results from the efforts reported here indicate that gene expression measurements can be robust. Furthermore, the results strongly support the notion that such measurements greatly compliment conventional end points and do not confound the results from traditional studies. All of this progress hinges on the standards of sample preparation and handling, microarray measurement, and data analysis that have been established by these consortia efforts. Those efforts that are ongoing clearly will move the field even further and provide the basis for focused, effective experiments conducted by individual organizations. For toxicogenomics, consortia have played, and will continue to play, a key role in the advancement of the science.

References

1. Schena, M., Shalon, D., Davis, R. W., and Brown, P. O. (1995) Quantitative monitoring of gene expression patterns with a complementary DNA microarray. *Science* **270**, 467–470.
2. DeRisi, J. L., Iyer, V. R., and Brown, P. O. (1997) Exploring the metabolic and genetic control of gene expression on a genomic scale. *Science* **278**, 680–686.
3. Anderson, N. L., Taylor, J., Hofmann, J. P., Esquer-Blasco, R., Swift, S., and Anderson, N. G. (1996) Simultaneous measurement of hundreds of liver proteins: application in assessment of liver function. *Toxicol. Pathol.* **24**, 72–76.
4. Kegelmeyer, A. E., Sprankle, C. S., Horesovsky, G. J., and Butterworth, B. E. (1997) Differential display identified changes in mRNA levels in regenerating livers from chloroform-treated mice. *Mol. Carcinog.* **20**, 288–297.
5. Rodi, C. P., Bunch, R. T., Curtiss, S. W., Kier, L. D., Cabonce, M. A., Davila, J. C., et al. (1999) Revolution through genomics in investigative and discovery toxicology. *Toxicol. Pathol.* **27**, 107–110.
6. Amundson, S. A., Bittner, M., Chen, Y., Trent, J., Meltzer, P., and Fornace, A. J. (1999) Fluorescent cDNA microarray hybridization reveals complexity and heterogeneity of cellular genotoxic stress responses. *Oncogene* **18**, 3666–3672.
7. Farr, S., and Dunn, R. T., 2nd. (1999) Concise review: gene expression applied to toxicology. *Toxicol. Sci.* **50**, 1–9.
8. Fornace, A. J., Amundson, S. A., Bittner, M., Myers, T. G., Meltzer, P., Weinstein, J. N., and Trent, J. (1999) The complexity of radiation stress responses: analysis by informatics and functional genomics approaches. *Gene. Expr.* **7**, 387–400.
9. Jelinsky, S. A., and Samson, L. D. (1999) Global response of Saccharomyces cerevisiae to an alkylating agent. *Proc. Natl. Acad. Sci. U. S. A.* **96**, 1486–1491.
10. Nuwaysir, E. F., Bittner, M., Trent, J., Barrett, J. C., and Afshari, C. A. (1999) Microarrays and toxicology: the advent of toxicogenomics. *Mol. Carcinog.* **24**, 153–159.
11. Lockhart, D. J., Dong, H., Byrne, M. C., Follettie, M. T., Gallo, M. V., Chee, M. S., et al. (1996) Expression monitoring by hybridization to high-density oligonucleotide arrays. *Nat. Biotechnol.* **14**, 1675–1680.
12. Shalon, D., Smith, S. J., and Brown, P. O. (1996) A DNA microarray system for analyzing complex DNA samples using two-color fluorescent probe hybridization. *Genome Res.* **6**, 639–645.
13. Stimpson, D. I., Cooley, P. W., Knepper, S. M., and Wallace, D. B. (1998) Parallel production of oligonucleotide arrays using membranes and reagent jet printing. *BioTechniques* **25**, 886–890.

14. Ermolaeva, O., Rastogi, M., Pruitt, K. D., Schuler, G. D., Bittner, M. L., Chen, Y., et al. (1998) Data management and analysis for gene expression arrays. *Nat. Genet.* **20**, 19–23.

15. Bassett, D. E., Eisen, M. B., and Boguski, M. S. (1999) Gene expression informatics—it's all in your mine. *Nat. Genet.* **21**, 51–55.

16. Zhang, M. Q. (1999) Large-scale gene expression data analysis: a new challenge to computational biologists. *Genome Res.* **9**, 681–688.

17. Brazma, A., and Vilo, J. (2000) Gene expression data analysis. *FEBS Lett.* **480**, 17–24.

18. ILSI. About ILSI. Available at http://www.ilsi.org/AboutILSI/.

19. ILSI. Application of Genomics to Mechanism-based Risk Assessment. Available at http://www.hesiglobal.org/Committees/TechnicalCommittees/Genomics/.

20. Parkinson, H., Kapushesky, M., Shojatalab, M., Abeygunawardena, N., Coulson, R., Farne, A., et al. (2007) ArrayExpress–a public database of microarray experiments and gene expression profiles. *Nucleic Acids Res.* **35**, D747–750.

21. ILSI HESI. Genomics Committee Status Report 2003. Available at http://www.hesiglobal.org/NR/rdonlyres/8F5B8FD4-B2C6–4DD0-A6D8-F5CBD60D8B13/0/CommitteeStatusReport.pdf.

22. Pennie, W., Pettit, S. D., and Lord, P. G. (2004) Toxicogenomics in risk assessment: an overview of an HESI collaborative research program. *Environ. Health Perspect.* **112**, 417–419.

23. Baker, V. A., Harries, H. M., Waring, J. F., Duggan, C. M., Ni, H. A., Jolly, R. A., et al. (2004) Clofibrate-induced gene expression changes in rat liver: a cross-laboratory analysis using membrane cDNA arrays. *Environ. Health Perspect.* **112**, 428–438.

24. Ulrich, R. G., Rockett, J. C., Gibson, G. G., and Pettit, S. D. (2004) Overview of an interlaboratory collaboration on evaluating the effects of model hepatotoxicants on hepatic gene expression. *Environ. Health Perspect.* **112**, 423–427.

25. Thompson, K. L., Afshari, C. A., Amin, R. P., Bertram, T. A., Car, B., Cunningham, M., et al. (2004) Identification of platform-independent gene expression markers of cisplatin nephrotoxicity. *Environ. Health Perspect.* **112**, 488–494.

26. Kramer, J. A., Pettit, S. D., Amin, R. P., Bertram, T. A., Car, B., Cunningham, M., et al. (2004) Overview on the application of transcription profiling using selected nephrotoxicants for toxicology assessment. *Environ. Health Perspect.* **112**, 460–464.

27. Amin, R. P., Vickers, A. E., Sistare, F., Thompson, K. L., Roman, R. J., Lawton, M., et al. (2004) Identification of putative gene based markers of renal toxicity. *Environ. Health Perspect.* **112**, 465–479.

28. Jolly, R. A., Goldstein, K. M., Wei, T., Gao, H., Chen, P., Huang, S., et al. (2005) Pooling samples within microarray studies: a comparative analysis of rat liver transcription response to prototypical toxicants. *Physiol. Genomics* **22**, 346–355.

29. Waring, J. F., Ulrich, R. G., Flint, N., Morfitt, D., Kalkuhl, A., Staedtler, F., et al. (2004) Interlaboratory evaluation of rat hepatic gene expression changes induced by methapyrilene. *Environ. Health Perspect.* **112**, 439–448.

30. Chu, T. M., Deng, S., Wolfinger, R., Paules, R. S., and Hamadeh, H. K. (2004) Cross-site comparison of gene expression data reveals high similarity. *Environ. Health Perspect.* **112**, 449–455.

31. Hamadeh, H. K., Knight, B. L., Haugen, A. C., Sieber, S., Amin, R. P., Bushel, P. R., et al. (2002) Methapyrilene toxicity: anchorage of pathologic observations to gene expression alterations. *Toxicol. Pathol.* **30**, 470–482.

32. Mattes, W. B. (2004) Annotation and cross-indexing of array elements on multiple platforms. *Environ. Health Perspect.* **112**, 506–510.

33. Goodsaid, F. M., Smith, R. J., and Rosenblum, I. Y. (2004) Quantitative PCR deconstruction of discrepancies between results reported by different hybridization platforms. *Environ. Health Perspect.* **112**, 456–460.

34. NIEHS. Introduction to the Toxicogenomics Research Consortium (TRC). Available at http://www.niehs.nih.gov/dert/trc/intro.htm.

35. Medlin, J. (2002) Toxicogenomics research consortium sails into uncharted waters. *Environ. Health Perspect.* **110**, A744–746.

36. Hosack, D. A., Dennis, G., Jr., Sherman, B. T., Lane, H. C., Lempicki, R. A., Yang, J., and Gao, W. (2003) Identifying biological themes within lists of genes with EASE. *Genome Biol.* **4**, R70.

37. Camon, E., Magrane, M., Barrell, D., Binns, D., Fleischmann, W., Kersey, P., et al. (2003) The Gene Ontology Annotation (GOA) project: implementation of GO in SWISS-PROT, TrEMBL, and InterPro. *Genome Res.* **13**, 662–672.

38. Tong, W., Lucas, A. B., Shippy, R., Fan, X., Fang, H., Hong, H., et al. (2006) Evaluation of external RNA controls for the assessment of microarray performance. *Nat. Biotechnol.* **24**, 1132–1139.

39. Shi, L., Reid, L. H., Jones, W. D., Shippy, R., Warrington, J. A., Baker, S. C., et al. (2006) The MicroArray Quality Control (MAQC) project shows inter- and intraplatform reproducibility of gene expression measurements. *Nat. Biotechnol.* **24**, 1151–1161.

40. Canales, R. D., Luo, Y., Willey, J. C., Austermiller, B., Barbacioru, C. C., Boysen, C., et al. (2006) Evaluation of DNA microarray results with quantitative gene expression platforms. *Nat. Biotechnol.* **24**, 1115–1122.

41. Patterson, T. A., Lobenhofer, E. K., Fulmer-Smentek, S. B., Collins, P. J., Chu, T. M., Bao, W., Fet al. (2006) Performance comparison of one-color and two-color platforms within the Microarray Quality Control (MAQC) project. *Nat. Biotechnol.* **24**, 1140–1150.

42. Guo, L., Lobenhofer, E. K., Wang, C., Shippy, R., Harris, S. C., Zhang, L., et al. (2006) Rat toxicogenomic study reveals analytical consistency across microarray platforms. *Nat. Biotechnol.* **24**, 1162–1169.

43. InnoMed PredTox. Available at http://www.innomed-predtox.com/.

44. FDA. (2006) Critical Path Opportunities Report and List (HHS, ed.). FDA, Bethesda, MD.

45. FDA. FDA and the Critical Path Institute Announce Predictive Safety Testing Consortium. Available at http://www.fda.gov/bbs/topics/news/2006/NEW01337.html.

46. Fielden, M. R., Brennan, R., and Gollub, J. (2007) A gene expression biomarker provides early prediction and mechanistic assessment of hepatic tumor induction by nongenotoxic chemicals. *Toxicol. Sci.* **99**, 90–100.

47. Nie, A. Y., McMillian, M., Parker, J. B., Leone, A., Bryant, S., Yieh, L., et al. (2006) Predictive toxicogenomics approaches reveal underlying molecular mechanisms of nongenotoxic carcinogenicity. *Mol. Carcinog.* **45**, 914–933.

12

Applications of Toxicogenomics to Nonclinical Drug Development: Regulatory Science Considerations

Frank D. Sistare and Joseph J. DeGeorge

Summary

Scientists in the pharmaceutical industry have ready access to samples from animal toxicology studies carefully designed to test the safety characteristics of a steady pipeline of agents advancing toward clinical testing. Applications of toxicogenomics to the evaluation of compounds could best be realized if this promising technology could be implemented in these studies fully anchored in the traditional study end points currently used to characterize phenotypic outcome and to support the safe conduct of clinical testing. Regulatory authorities worldwide have declared their support for toxicogenomics and related technological tools to positively impact drug development, and guidance has been published. However, applications of exploratory "omics" technologies to compounds undergoing safety testing remain inhibited due to two core data submission responsibility implications and ambiguities: *(1)* constraints arising from continual literature surveillance and data reanalysis burdens, under the shadow of looming subsequent reporting requirements to regulatory authorities as gene expression end points loosely linked to safety gain attention in the published literature, and *(2)* ambiguities in interpretation of validation stature remain between exploratory, probable valid, and known valid safety biomarkers. A proposal is offered to address these regulatory implementation barriers to open access for exploring this technology in prospective drug development animal toxicology studies.

Key Words: biomarker; critical path; drug development; drug safety testing; genomics; guidance; investigative toxicology; lead optimization; metabolomics; proteomics; qualification; regulation; toxicogenomics; validation.

From: *Methods in Molecular Biology, vol. 460: Essential Concepts in Toxicogenomics*
Edited by: D. L. Mendrick and W. B. Mattes © Humana Press, Totowa, NJ

1. Introduction

Since the introduction 12 years ago of genome-scale measurement technologies that can monitor thousands of changes in transcripts between samples *(1)*, routine applications of such approaches to drug development safety assessment groups have not flourished at the pace that many had anticipated. Two fundamental strategic business applications of toxicogenomics to toxicologic assessments in drug development decision making are slowly emerging: (1) advancing or measuring qualified biomarkers, and (2) generating testable hypotheses using exploratory end points. The first is centered on the development and subsequent recognition of altered changes in gene expression in a tissue as new legitimate qualified *biomarkers* contextually appropriate for some level of decision making. The interpretation of these biomarkers would be well grounded in results from robust testing paradigms and a firm scientific understanding from which conclusive interpretations regarding toxicologic study outcome can be drawn. The other application of these technologies is far more speculative in nature. This second fundamental application relies on reasonable associations of alterations of transcript changes with certain outcomes but are poorly established and loosely drawn. The utility of the interpretation of these latter exploratory markers is to formulate reasonable and testable hypotheses that may come from a small number of internal or even published experiments, or from a perceived role expected of their protein products in cellular processes or biochemical pathways that have been collected into software programs. The goal, therefore, of the second application is to *generate testable hypotheses*. Existing regulations, regulatory guidance, and defined data-reporting requirements frame and unfortunately limit business opportunities and strategies to implement this technology to real-time drug development for both advancing and measuring qualified biomarkers and for generating testable hypotheses.

2. Regulatory Guidance Development to Support Expected Regulatory Applications

Regulatory authorities have declared the need to collaborate to expand opportunities that encourage applications of "omics" technologies and thereby accelerate safety biomarker development for improved and accelerated toxicologic risk assessment. There is regulatory optimism that toxicogenomics will yield greater insight and be more predictive than current toxicology testing methods. To facilitate the availability of this technology, regulatory authorities have developed and published guidance to clarify data submission requirements so that these exploratory technologies would be seamlessly integrated into drug development and regulatory review practice.

2.1. Encouragement by Regulatory Authorities

There is a general belief that broad knowledge of drug-associated alterations in global gene expression, protein expression, protein modifications, and protein redistribution will enhance the efficiency and pace of drug development and therefore benefit the practice of clinical medicine and promote human health *(2)*. Food and Drug Administration (FDA) authors, in their publication, "Innovation or Stagnation: Challenge and Opportunity on the Critical path to New Medical Products," point furthermore to a pressing need for such translational critical path research. "The goal of critical path research is to develop new, publicly available scientific and technical tools—including assays, standards, computer modeling techniques, biomarkers, and clinical trial endpoints—that make the development process itself more efficient and effective and more likely to result in safe products that benefit patients. Such tools will make it easier to identify earlier in the process those products that do not hold promise, thus reducing time and resource investments, and facilitating the process for development of medical products that hold the most promise for patients" *(3)*.

Similarly, European Medicines Agency (EMEA) authors have noted in their Road Map to 2010 that "one of the key issues to be addressed by the international regulatory environment will be its ability to prepare adequately for the introduction of new technologies, from a scientific, legal and regulatory perspective" *(4)*.

The EMEA/CHMP (Committee for Medicinal Products for Human Use) Think Tank report issued in 2007 states further that "[t]he qualification of biomarkers and surrogate endpoints was brought up by all major pharmaceutical companies as one of their most important topics. To gain efficiency, the collaboration between individual companies, as well as between industry, academia, and regulatory agencies is acknowledged. Industry would particularly welcome harmonisation between the EMEA and the FDA." The think-tank group recommends that the EMEA "... commits itself to this global collaboration through the establishment of regular discussion fora and the implementation of global scientific guidance" *(5)*.

That drug development needs improvement is apparent from industry surveys indicating that fewer than 10% of drug candidates evolving through the complex nonclinical developmental pipeline to begin Phase I clinical trials ultimately receives marketing approval with unacceptable toxicities and safety concerns cited to account for at least 20% of the failures *(6)*. One interpretation for these inefficiencies is that the data now being generated in early animal toxicology studies are insufficiently informative to address and accurately "predict" outcomes and margins of animal studies of longer duration. There is belief that these new "omic" data could be used to predict an unfavorable exposure based balance between human efficacy and safety concerns that will surface from later clinical studies. It is reasoned that, perhaps if more and better data were generated earlier from end points that were more informative and predictive of human outcome, and if more were known of early accessible phenotypic bridging biomarkers of response that could detect the differences that may exist in individual human sensitivities, then better drugs could be developed quicker as less favorable candidates are more rapidly identified and discarded. The FDA authors point further to specific opportunities: "Proteomic and toxicogenomic approaches may ultimately provide sensitive and predictive safety assessment techniques; however, their application to safety assessment is in early stages and needs to be expanded" *(3)*.

Regulatory scientists from regions outside of Europe and the United States *(7)* have also expressed the similar belief that "... changes in gene/protein can be used to predict toxicity or the toxicological outcomes following exposure to drugs or xenobiotics.... With carefully selected genes, gene expression data can provide valuable information to be incorporated into risk assessment." "[D]ifferential gene expression ... holds the promise to contribute significantly to risk assessment, particularly in those target areas where the predictivity of the current nonclinical studies is low." But the authors recognize that "for regulators to take these data into consideration for risk assessment, it is important that gene expression assays are validated."

2.2. Development of Regulatory Guidance to Support Toxicogenomics Data Submissions

Codified regulations in the United States mandate that industry sponsors provide adequate information about the pharmacologic and toxicologic properties of a drug, as described in 21 CFR 312 *(8)*. "A sponsor who intends to conduct a clinical investigation subject to this part shall submit an 'Investigational New Drug Application' (IND) including ... pharmacology and toxicology information, ... adequate information about pharmacological and toxicological studies of the drug involving laboratory animals or in vitro, on the basis of which the sponsor has concluded that it is reasonably safe to conduct the proposed clinical investigations.... As drug development proceeds, the sponsor is required to submit informational amendments, as appropriate, with additional information pertinent to safety ... for each toxicology study that is intended primarily to support the safety of the proposed clinical investigation, a full tabulation of data suitable for detailed review."

The appropriate presentation of pharmacology and toxicology data is further described in an FDA guidance document *(9)*. For pharmacology studies, a summary report without individual animal records is usually sufficient for presentation to the FDA's Center for Drug Evaluation and Research (CDER). In contrast, as noted above, regulations require a full tabulation of the toxicologic effects of the drug in animals and *in vitro*. The CFR and FDA guidance specify therefore that study reports containing a full tabulation of the data suitable for detailed review are to be made available for each study. For this purpose, final, fully quality-assured individual study reports are usually available at the time of submission, or within 120 days of submission of unaudited draft reports. In addition, as required under 21 CFR 312.23 *(8)*, a statement that nonclinical laboratory studies subject to good laboratory practice regulations were conducted in compliance with good laboratory practices (GLPs), or documentation for noncompliance and the impact of noncompliance on study results and interpretations, should also be included in study reports.

The FDA's Center for Drug Evaluation and Research recognizes that there is some interpretation as to the extent that new nonclinical genomic data do or do not need to be submitted to fulfill this obligation *(10)*. Submission of data that may have little relevance to safety decisions will burden both regulatory agencies and sponsors with studies that will add no value to the regulatory oversight and review. These types of studies may more appropriately be classified as discovery or research studies rather than as pharmacology or toxicology studies. This may be especially applicable to genome-wide expression profiling, as the quantity of data can rapidly consume human and information resources at regulatory agencies that could arguably be better spent. Regulatory authorities have recognized the need on their part to clarify, in this context, exactly what general data format detail could be followed by a sponsor desiring to submit such data. U.S. FDA *(11)* and European *(12)* guidances have been published, for example, to describe the general proposed format and content for how such data might be submitted by a drug sponsor and further detailed European *(13)* and FDA guidance *(14)* specific to toxicogenomics are expected soon.

However, the issues of (1) which drug testing studies the submission of such data would be mandated, (2) whether all or only certain well understood components of the data need be submitted, and (3) the duration of responsibility for reanalyses of data from emerging biomarkers surfaced quickly and remain areas of concern. Does it matter if the microarray gene expression data are generated as a component of an early pharmacology rather than a GLP toxicology experiment? What does this mean for gene expression data—a mean and standard deviation for each gene, data for just significantly altered genes, just genes whose biological function and expression patterns can be explained by the investigator? As publications of exploratory gene expression biomarkers appear over time in the published literature, what are a drug sponsor's responsibilities to review and reanalyze their genomic data generated years earlier on the test agent that may have advanced to Phase III clinical testing?

The concept of *safe harbor* for genomics data was introduced in a May 2002 workshop sponsored by Pharmaceutical Research and Manufacturers of America (PhRMA) and FDA *(15)* for genome-based data. Doubt that pharmacogenomic and related global technologies are sufficiently mature to provide data useful for safety or efficacy decisions was and continues to be expressed. A significant concern is that individual review scientists in key positions in regulatory bodies such as the FDA Center for Drug Evaluation and Research may take premature regulatory action on components of toxicogenomic study data based on early publications that confuses rather than enlightens understanding of drug safety and efficacy. Regulatory agencies and knowledgeable investigators believe, however, that these global technologies have the underlying potential to positively impact drug development through clinical and nonclinical drug development efficiencies. Because of the perceived anticipated benefits of the technology and the concern on the part of the regulated industry, the concept of safe harbor was introduced.

The term *safe harbor* was used to describe a process in which genomic data could be generated on lead compounds and submitted under an active IND but would not carry regulatory significance unless the end point had already been demonstrated to have sufficient validity. Under this concept, toxicogenomic studies would not undergo formal regulatory review until more was known about the validity of the technology used and the appropriate interpretation of the data, even though the samples could derive from GLP toxicology studies conducted by industry scientists on compounds at the IND stage. The data would be submitted to the agency without regulatory impact unless a certain degree of validation had been conducted. To use the example of exploratory genomic-based data often cited that might have tenuous or uncertain interpretation is the measured activation or overexpression of an oncogene in an expression profiling experiment, possibly leading to adverse regulatory action. If regulatory reviewers interpreted such a finding to raise concerns over such an example nonmonitorable condition as tumorigenesis, a heightened level of concern is raised among industry scientists given that it could signify a tumorigen or a completely unrelated response. Such adverse regulatory actions might be envisioned to include placing the product on clinical hold, informing the institutional review board and modifying patient consent, requiring additional or earlier completion of nonclinical studies to support any further clinical testing, or perhaps adding clinical monitoring for such end points that would not be standard for a Phase I or II trial nor scientifically established.

Such a measured change in oncogene expression could have no legitimate interpretation without extensive experimental validation or qualification *(16)*. Extensive research is needed to first establish the relationship between drug-induced carcinogenesis and altered expression of specific oncogenes, tumor suppressor genes, regulators of genomic stability, and DNA repair proteins for example. If the sensitivity, specificity, positive predictivity, and negative predictivity of the linkage of changes in sets or even individual critical gene expression changes to carcinogenesis can be established with sufficient rigor, such end points could be extremely useful to all concerned with safe drug development and regulation. But only after these linkages are established in a scientifically robust manner and consensus has been achieved would invoking regulatory intervention be judicious. Until such experimental validation or qualification is established, "pertinence to safety" remains speculative and dubious.

The regulatory boundary between speculation and "pertinence" remains unclear, however. According to FDA guidance, a probable valid biomarker with specific regulatory reporting requirements may earn its stature from "highly suggestive" data sets that "may not be conclusive" *(11)*. To continue the analogy using predictors of tumorigenesis for example, a peer-reviewed publication declares that "... genes known to be induced by the transcription factor and tumor suppressor p53 were upregulated by ... four carcinogens, including the apoptosis inducer BAX, the cyclin-dependent kinase inhibitor p21, the cell growth regulatory gene 11, the B-cell translocation gene 2, cyclin G1, and the ubiquintin E3 ligase MDM2, ... to some extent also NADPH quinone oxidore-ductase ... suggesting activation of the ARE pathway" *(17)*. In a separate publication, other authors state that "... MDM2 induction ... might be a critical step in prostate growth regulation., as MDM2 induces the nuclear export and degradation of p53 ... leading to a number of events including reduced transcription of cell cycle inhibitors Rb, ... [BTG2/TIS21/PC3, ID-2]... reduced IGF-BP3 (enhanced IGF-1 signaling)" *(18)*. A third manuscript impli-cates some of the same transcripts and pathways stating their success at having "... identified several candidate molecular markers of rodent nongenotoxic carcinogenicity, including transforming growth factor beta stimulated clone 22 (TSC-22) and NAD(P)H cytochrome P450 oxidoreductase (CYP-R)" *(19)*. To industry scientists eager to understand and apply the technology to GLP toxicology study samples, the potential regulatory implications for generating and either submitting or not submitting such data points are perilous to a devel-opment program when a probable valid biomarker is defined using "highly suggestive data that may not be conclusive."

Thus, without specifically addressing the term *safe harbor* in its guidance for industry on pharmacogenomic data submissions published on March 22, 2005 *(11)*, the FDA acknowledges that "... pharmacogenomics is rapidly evolving... [and] ... results may not be well-established scientifically to be suitable for regulatory decision-making." The guidance is being issued to "facilitate the use of pharmacogenomic data in drug development" ... and ... "facilitate identification of predictors of safety, effectiveness, or toxicity." The guidance acknowledges that, "at the current time, most pharmacogenomic data are of an exploratory or research nature and FDA regulations do not require that these data be submitted to an IND ... ." The guidance therefore encourages voluntary submission of genomic data that is otherwise not required for all to learn and gain greater experience. The guidance recognizes that there will exist "biomarkers for which validity is not established," "known valid," and "probable valid" biomarkers.

According to the guidance *(11)*, a biomarker is a "known valid biomarker if (1) it is measured in an analytical test system with well-established performance characteristics, and (2) there is an established scientific framework or body of evidence that elucidates the physiologic, pharmacologic, toxicologic, or clinical significance of the test results." A probable valid biomarker is a biomarker that "is measured in an analytical test system with well-established performance characteristics and for which there is a scientific framework or body of evidence that appears to elucidate the physiologic, toxicologic, pharmacologic, or clinical significance of the test results. A probable valid biomarker may not have reached the status of a known valid marker because, for example, of any one of the following reasons:

- The data elucidating its significance may have been generated within a single company and may not be available for public scientific scrutiny.
- The data elucidating its significance, although highly suggestive, may not be conclusive.
- Independent verification of the results may not have occurred."

This categorization conveys the philosophy that there is an inherent shared responsibility on the part of both the drug development industry and regulatory authorities to share knowledge of those end points for which internal nonpublic data would support the context and validity of the end point as a *bona fide* meaningful biomarker with clearly interpretable strengths and limitations—the probable valid biomarker. The guidance document indicates essentially that two factors drive whether genomic data submission is required for an IND filing application: (1) the study purpose and (2) the level of confidence in the qualification of the biomarker (**Table 1**). Submission of such genomic data to an IND is not required if:

1. "Information is from exploratory studies." Early non-GLP dose range finding and compound selection studies that are used for internal decision making by the industry and are *not* used to establish the safe conduct of a clinical study would fall into this category.
2. Or "results are from test systems where the validity of the biomarker is not established." Those end points sufficiently qualified experimentally to be classified as probable and known valid biomarkers will rightfully demand a higher level of regulatory scrutiny when generated from a GLP study than would the exploratory end points for which no validity has been established.

Table 1
Two Factors Drive IND Data Submission

	Level of qualification		
Study purpose	Low (Validity not established)	Medium (Probable valid)	High (Known valid)
1. GLP study with potential for data with regulatory impact			
a. Used by sponsor in decision to support the safety of a clinical trial	Voluntary*	Full report	Full report
b. Not used to assess prognosis of animal findings (e.g., explore mechanism)	Voluntary*	Voluntary*	Abbreviated report
2. Non-GLP exploratory study for internal decisions	Voluntary	Voluntary	Voluntary

*If additional information becomes available that triggers the requirements for submission under §§ 312, 314, or 601, the sponsor must submit the data...

3. General Concerns of Drug Developers for Generating Exploratory Genomic Data from GLP Toxicology Study Samples

Within the operational boundaries established by this regulatory guidance, there remain two general concerns that drug development scientists face when considering opportunities to generate toxicogenomic data in a regulatory GLP animal toxicology study that will be used for safety assessment purposes:

3.1. Resources to Support a Moving Target

For industry scientists seeking to conduct such studies and to appropriately interpret their reporting responsibilities and include the requisite data in their regulatory filings, this guidance would appear to certainly open the door to generate exploratory genomic data from internal decision-making exploratory non-GLP studies, and also even at GLP toxicology testing stages with little concern for disagreement over regulatory interpretation as only the probable and known valid biomarker components of such broad microarray data sets would require submission to the IND. The guidance document indicates that exploratory data generated even from GLP toxicology studies could form the basis for a voluntary genomic data submission to an interdisciplinary regulatory review group and not constitute a required data component for an IND application to a review division, thereby allaying concerns of nondisclosure accusations while maintaining some comfortable separation from the individual product review scientific staff. It is reasoned that the submission of nonclinical pharmacogenomic data to the agency would allow reviewers and sponsors to gain experience with the data and data-sharing process and allow confidence to grow in the appropriate integration of toxicogenomic data into the regulatory study and review paradigm. As the agency gains experience with nonclinical toxicogenomic data, and appropriate interpretation of gene expression data components are embraced by the scientific community, a new approach to data review could then become incorporated into CDER's good review practices *(20)*.

However, the guidance document *(11)* acknowledges that advances in bioinformatics and toxicology could lead to a higher level of understanding over time for end points that would clearly be classified merely as exploratory at the time the data are generated. As indicated above, 21 CFR 312 stipulates that, "as drug development proceeds, the sponsor is required to submit informational amendments, as appropriate, with additional information pertinent to safety." The CDER has added clarification in the 2005 guidance *(11)*, indicating that "if additional information becomes available that triggers the requirements for submission under §§ 312, 314, or 601, the sponsor must submit the data to the relevant application and should follow the appropriate algorithm." This clearly keeps industry scientists outside of a "safe harbor" and would put them in the unenviable position of having to constantly review and revisit the evolving analytical and biological validity of all microarray data points from all tested compounds over their lifetimes of development and then submit them to the regulatory authorities under the IND. While there may be little understanding of the genomic data when they are first generated, final reporting for marketing applications for all studies of each development program will be completed many years later as additional longer animal toxicology and human clinical studies are completed. There are requirements all during that time for regularly updating regulatory authorities with any new information to the IND application that are "pertinent to safety."

Anonymous Expressed Sequence Tags (EST) probe sequences on current-generation microarrays will become associated over time with established genome transcripts, and new claims are likely to appear in the peer-reviewed literature over this time frame that will link alterations in transcript abundance with claims of predictions for deleterious safety outcomes. If these are to be considered probable valid biomarkers with pertinence to safety, then this places a high burden on both drug development and regulatory review scientists to exercise the vigilance that could be expected and required to avoid accusations of nondisclosure and fraud or dereliction. Otherwise, the sponsor might be viewed as intentionally not reporting data that could have an impact on the interpretation of a study, or the regulatory reviewer might be considered derelict in their responsibilities for not diligently maintaining full awareness of pertinent public information affecting the product. Regular systematic reassessment and scrutiny over old data, mapping probesets across microarray platform version updates and data sets, and regular surveillance against the published literature to support scores of concurrent development programs all moving forward simultaneously may be necessary to avoid nondisclosure or dereliction concerns. This is an unattractive option to devote resources to, as longer toxicology and human clinical studies would supersede with definitive outcome data any claims of possible predictable genomic data from shorter-duration animal studies. Public acknowledgment of this by regulatory authorities is needed.

The alternatives for industry scientists are either to assume the unacceptable risk liabilities associated with the nondisclosure of what may be deemed important "information pertinent to safety" or not conduct such genomic assessments. The option of routinely submitting all data to the traditional IND review process nullifies the intent and value of the guidance as it was established. And as mentioned above, it would furthermore only shift the continuous data reanalysis and monitoring responsibility to an already overburdened CDER review process. Current staffing shortages for the evaluation of existing regulatory submissions would be exacerbated. These IND reporting concerns are limiting willingness by industry sponsors to generate such toxicogenomic data sets on the continuum of progressively longer-duration GLP animal toxicology studies that could truly elucidate the technology's strengths and limitations.

For marketing a new drug, under 21 CFR 314.50 *(21)*, "the application is required to contain reports of all investigations of the drug product sponsored by the applicant, and all other information about the drug pertinent to an evaluation of the application that is received or otherwise obtained by the applicant from any source . . . ." By the time a marketing application is prepared for regulatory submission, all of the studies that could determine and lay to rest any ambiguities of signals inherent in toxicogenomic data from earlier and shorter animal studies would be clearly superseded by the traditional data from subsequent longer-term animal toxicology studies and human clinical trials. There is no concern by industry sponsors therefore in the requisite submission of reports containing toxicogenomic data at this point in development, in accordance with the published guidance (**Table 2**).

Table 2
NDA Data Submission Guidance

	Level of qualification		
Study purpose	Low (Validity not established)	Medium (Probable valid)	High (Known valid)
1. GLP study with potential for data with regulatory impact			
a. Used by sponsor to support a safety claim	Synopsis	Abbreviated report	Full report*
b. Not used to assess prognosis of animal findings	Synopsis	Abbreviated report	Abbreviated report
2. Non-GLP exploratory study for internal decisions	Synopsis	Abbreviated report	Abbreviated report

* 314.50 CFR: "... the [NDA] application is required to contain reports of all investigations of the drug product sponsored by the applicant."

There is a potential solution and a path forward to break this impasse and alleviate concerns surrounding generation of toxicogenomic data from IND enabling GLP toxicology studies. Regulatory authorities could promote and advance further evolution of toxicogenomics practice by acknowledging similarly that during IND stages of compound development, if a definitive study of longer duration were complete, then toxicogenomic data from an earlier shorter-term GLP study need not be continually subjected to further scrutiny and reanalysis but could instead be laid to rest until the NDA filing stages (**Table 3**).

3.2. Expectations of Alternative Interpretations of Ambiguous Data

As discussed above, industry scientists could readily avoid nondisclosure concerns from not submitting data simply by submitting all of the genomic data generated from samples off GLP toxicology studies. Besides serving to transfer much of the burden of searching for "potential signals" to regulatory authorities who are not sufficiently staffed to cope with such enormous volumes of data,

Table 3
Proposed Solution for IND Submissions

	Level of qualification	
Study purpose	Low (Not qualified)	High (Qualified for regulatory use)
1. GLP study with potential for data with regulatory impact		
a. Used by sponsor in decision to support the safety of a clinical trial	Voluntary*	Full report
b. Not used to assess prognosis of animal findings (e.g., explore mechanism)	Voluntary*	Full report
2. Non-GLP exploratory study for internal decisions	Voluntary	Voluntary

* If additional information becomes available prior to completion of the pivotal subsequent GLP toxicology study that triggers the requirements for submission under §§ 312, 314, or 601, the sponsor must submit.

this option also creates angst resulting from the ambiguity of interpretation inherent in these emergent "potential signals."

Applications of global measures of gene expression alterations in various tissues of animals after drug exposures have been used at several levels of analyses: (a) global patterns of expression changes of whole arrays of thousands of genes have been used to define and distinguish drug class effects; (b) changes in subsets of genes associated with specific biochemical pathways or processes have been used as "signatures" to suggest explanations of certain intended or unintended drug activities; (c) measured changes in the expression of select individual transcripts are sought as highly predictive individual biomarkers of drug effect. Each level of analyses—global patterns, signatures, or the individual biomarker transcripts—could be valuable to improve understanding of drug actions and provide some confidence in predicting longer-term study outcomes. It is important to recognize, however, that gene products can have many biological roles, many if not most of which have not been defined. Because of this, there are ambiguities in interpretation, and if this ambiguity is not fully acknowledged at the drug development/regulatory interface, there is high potential for impeding incorporation of the new technology into common drug development practice.

Safety assessment GLP study requirements place a heavy focus and expectation on careful documentation, careful compliance to established operating procedures, and accurate and precise interpretation of study results with little room for speculation and data ambiguities. As currently constrained, such GLP studies are therefore not attractive sources for samples that could be used to implement and prospectively "test drive" innovative technology solutions with creatively speculative interpretations that could generate many testable hypotheses and eventually drive scientific evolution. The shear numbers of data points from each study animal, of the order thousands compared with the hundreds of traditional study data end points that have endured over decades of practice, present a formidable consensus-building interpretation challenge. Whereas some subsets of data will certainly evolve into known valid biomarker stature over time, all of the data from a microarray experiment are likely to remain impossible to interpret with confidence and mutual agreement for quite a long time. Different readers of the data will focus on different subsets of changes and create many different speculative hypotheses. Drug development scientists and managers remain concerned of potential further derailment of tight drug development timelines as regulatory reviewers would also be expected to raise their own concerns using their own and likely different algorithms and speculative interpretations of the same data.

If understanding or convincing evaluations of the linkage of gene expression or protein alterations to biological outcome have truly not matured, then the premature use of "omic" data to influence any decision-making step would be ill advised for both industry scientists and regulatory reviewers. On the other hand, as institutional experiences and the peer-reviewed scientific literature and databases grow, understanding in some specific areas involving small, more fully evaluated subsets of the data will grow faster than others, and components of huge data sets will truly become informative. Such bioinformatic limitations inherent in convincingly and accurately explaining the biological implications of all of the data can lead to the generation of many and diverse hypotheses, each of which requiring a prioritization decision to determine the best follow-up research strategy. This ambiguity of interpretation is restricting "omics" to exploratory research projects and is impeding full prospective applications of these technologies to investigate the toxicity potential of novel compounds intended for clinical development and regulatory review and oversight.

Concerns are magnified when the uncertainty of the biological interpretations by the data generator are considered together with the uncertainty over how the data will be interpreted and used by regulatory reviewers responsible for approval of critical stages of product investigation. The poorly understood exploratory transcriptional changes are expected to present a "Pandora's box" of hypotheses generating data to the regulatory reviewer charged with assuring the continued safety of drugs under clinical investigation, as well as with the responsibility of informing and protecting the safety of clinical trial participants. We are simply not able at the present time to reliably interpret the human health implications of the full extent of these exploratory data streams. Applications of these technologies to develop and enhance collective under-standing of exploratory biomarkers are being held hostage therefore due to constraints arising from continual surveillance and reanalysis burdens, and the shadow of looming subsequent reporting requirements to regulatory authorities. There is a critical need to strike the proper balance that enables the application of toxicogenomics measurements of end points that are not fully understood without adding risk to the drug development path for a new therapeutic agent.

Our understanding of gene expression responses has not matured as to which alterations represent reporters of intended and unintended pharmacology, which can be relied upon to predict later-appearing toxicities, and which represent healthy compensatory responses to xenobiotic exposures that act to normalize tissue function. It is also unclear to most researchers how the majority of these changes relate to biological outcome, and what levels of change equate to intolerable irreversibly compromised tissue function. Elements of a gene expression profile reflecting toxicology are difficult to distinguish from those reflecting pharmacologic effects induced by drug treatment. Food restriction was shown, for example, to induce subsets of genes that overlapped with the gene induction profile seen with chronic dietary exposure to diethylhexylph-thalate, which reduced food consumption in dosed mice *(22)*. Such data point out that healthy, physiologic, compensatory, and adaptive responses may not be easily distinguishable from causative, contributory, or corollary toxicologic responses. Recognition and enablement is needed from regulatory authorities to capture the tremendous value that would be gained if a greater freedom to operate were extended to drug development scientists to allow the prospective use of GLP toxicology study samples and to link data from the continuum of studies from early non-GLP exploratory studies to first-in-human enabling short-term GLP toxicology studies, to GLP subchronic and chronic animal toxicology studies, and ultimately to human clinical trial outcomes. It has been stated that to date the agency has seen few submissions that include toxicoge-nomic gene expression data from animal toxicology studies. The few that were submitted were generated as part of the voluntary genomic data submission process established to encourage mutual understanding of the technology.

To advance the development of this potentially useful technology, further guidance clarification and a transparent public-private partnership, as will be discussed below, are imperative in order to develop and communicate consensus as to when a signal can clearly meet the qualification stature of a biomarker sufficiently free of ambiguity that is fit for regulatory purposes involving decisions with safety implications. The proposed path forward (**Table 3**) reasons that if the measurement of a new end point has reached the stature of biomarker qualification up front that is appropriate for regulatory decision making and pertinent to safety, then it is considered qualified and the data should be carefully considered and integrated by industry scientists and reported to the IND for regulatory review. The biomarker may be qualified based either on data and experience that the sponsor has invested in privately or because of a transparent public acceptance and communication process that regulatory scientists are actively and responsibly engaged in. If the biomarker has not reached such a qualification status, then it remains exploratory and is inappropriate for regulatory decision making by review scientists or for making safety claims by industry scientists and therefore could be reserved for submission at the time of the NDA filing and would not be a required submission to the IND. The concept of a probable valid biomarker with some but perhaps insufficient qualification data should be carefully reconsidered for future guidance.

4. Expected Earliest Business Implementations of Nonclinical Toxicogenomics to Drug Development

Implementation of each business strategy to incorporate or to not generate genomic data as part of a drug development plan will be framed and limited by existing regulation and new regulatory guidance. Under the current regulatory framework, then, given the ambiguities in biomarker qualification stature and the burden of responsibilities for continual reanalysis of older data, the earliest impact and safest implementation strategy for toxicogenomics on drug development is occurring in two general areas: lead optimization and targeted investigative toxicology.

4.1. Toxicogenomics and Lead Optimization

In the near term, it is anticipated that the most common implementation of gene expression data may be felt as sponsors generate higher confidence in their knowledge of small subsets of the data, to incorporate the technology into lead optimization investigations to screen more rapidly against expected toxicology liabilities. Flowing from experience with such applications are expectations that there would be fewer surprises of low-margin adverse drug actions that impede later and more resource intensive GLP toxicology testing stages of drug development. As a result of such lead optimization studies, sponsors may be prompted by their own data to direct their earliest toxicology investigations at very specific critical questions. The guidance is clear that submission to the agency of such data in an IND from lead optimization studies to select compounds for development, per se, would not be required, but submission to regulatory authorities may be appropriate for voluntary motives as industry scientists begin to detect that such data subsets constitute possible biomarker characteristics. A major distinction exists, however, between a biomarker that may not be qualified for safety assessment use during regulated stages of drug development versus that which may be qualified for lead optimization internal decision making stages where significant false-positive rates may be acceptable given the low investment accumulated at that point.

4.2. Toxicogenomics and Targeted Investigative Toxicology

There are doubts that genomic data are sufficiently robust and sufficiently well understood to convincingly identify most safety concerns. Therefore, such study end points are not expected at this time to replace or reduce traditional toxicology end points or to modify traditional determinants of clinical starting doses. Time may prove otherwise, but it is likely that they will be most useful to identify newer and improved biomarkers whose context can be well appreciated once qualified and used in more traditional ways. Because there is also apprehension that regulatory reviewers may act prematurely to respond to genomics data in a manner that could adversely impact a drug's development, routine generation and regulatory submissions of toxicogenomic data are not expected from industry on compounds under active development. However, because confidence will grow in certain components of genomic data, there is likely to be growth in regulatory applications of adjunct data submissions directed at interpreting traditional nonclinical findings with potential clinical safety implications. Such investigational toxicology applications may be directed for example at attempting to understand whether mechanisms of toxicity may be species specific and irrelevant to humans or to identify mechanism-based bridging biomarkers that could be used specifically to investigate human relevance of animal findings. Nonclinical toxicogenomic studies might serve, then, to identify a robust focused subset of transcripts for additional study by more traditional methods (e.g., quantitative RT-PCR, RNAse protection assays, Northern blots, etc.) or of their protein products along with the development of a convincing data set establishing the utility of the biomarkers as qualified for making safety decisions. The current guidance is clear that probable and known valid biomarker status is expected if such data sets are to influence regulatory decisions in a weight-of-evidence argument (**Table 1**).

5. The Need for a Public-Private Partnership and a Transparent Process to Qualify Toxicogenomic Biomarkers for Regulatory Applications

There is a need to clarify when an exploratory biomarker crosses the threshold to a more extensively qualified biomarker and, after it does, to define what a sponsor's responsibilities will be to implement such end points into all future development strategies. As reviewed recently *(16,23)*, it must be acknowledged that the qualification of biomarkers is for specific defined uses and would be appropriate on a case-by-case situation and not for all situations. Further, it is becoming evident that there is a need for a public-private partnership to establish a procedure to qualify new exploratory biomarkers as context specified qualified safety biomarkers appropriate for regulatory decision making *(16,24)*. To meet that challenge, it will be important to further establish certain details, for example even the magnitude of changes in the qualified biomarkers of toxicity that constitute an "action level" beyond a recognized and accepted threshold of normal variation.

Nonclinical toxicogenomic experiments will, in general, be expected to become useful to the regulatory evaluation process when they convincingly expand the foundation of biological knowledge that could be used to modify expectations of clinical outcome. Nonclinical toxicogenomic studies could serve as the starting point for identifying intended and unintended pharmacology that will be reflected in gene expression changes that may be useful predictors of an animal's response, and such questions can be systematically investigated. However, claims that transcript changes in an animal species that may predict rare human toxicities even when the same animal species does not present with those clinical toxicities will be extremely difficult if not impossible to formally qualify or validate. Owing to the nature of drug development, there are unlikely sufficient compounds or capabilities to retrospectively test and to then prospectively confirm such claims.

The hope for realizing the predictive potential of toxicogenomics is based on the hypothesis that distinct gene expression profiles can be derived for toxicants that act via different mechanisms or modes of action to be recognized at times that precede a fully manifest tissue pathology. This has been demonstrated in several publications where blinded chemicals were correctly classified by broad toxicologic category or the type of histopathology induced at later times based on the similarity of their gene expression patterns to profiles of compounds with well-characterized toxicologies *(25–29)*. It is one matter to identify gene expression changes through class comparisons that appear to be both biologically and statistically significantly associated with a drug exposure class effect and to identify a "chemical fingerprint" in a single laboratory using one type of microarray platform. Analyses of those changes may be useful for chemical fingerprinting and hypotheses generation each of which may be of some value to an individual sponsor for lead optimization and compound selection as discussed above. Combined supervised and unsupervised clustering approaches to identify critical differences between members of the same class using class discovery strategies may also be useful to a sponsor for a more fine-tuned iterative process of repeat rounds of careful selection using further modified and improved compounds. But it is another matter to derive a highly valuable reduced set of genes whose drug-induced expression changes will universally predict a toxicologic outcome. The ability to identify and link specific sets of gene expression changes with the onset of different mechanisms of injury or the appearance of different pathologic processes may be goals of this technology for many investigators. To be meaningful, the term *predict* should imply that the data have survived a rigorous validation or qualification. For expanded sets of gene expression alterations to become routinely incorporated into scientifically based practices of drug development and regulation, this qualification or validation assessment would best be conducted in a transparent manner open to public scrutiny, criticism, and acceptance. Again, it is one feat to accomplish this in one laboratory under a defined set of carefully and rigorously controlled protocols on a single platform. And it is quite another feat to demonstrate that the same protein and gene expression changes can be rigorously associated with the relevant phenotypic outcome across many studies, many drug classes, different strains, numerous laboratories, and multiple analytical platforms. Proprietary database vendors have made claims to their knowledge of validated predictive sets of gene expression alterations relevant to toxicities, but it is fair to say that no robust cross-validated toxicity gene expression set currently exists in the public domain at this time. Independent analytical validation followed by biological qualification across different classes of compounds and several laboratories would be needed.

Coordinated consensus-building efforts are needed to integrate improved toxicity biomarkers into common regulatory practice. Robust and analytically valid assays for candidate biomarkers are needed that can then be applied to collaborative research programs to extensively evaluate strengths and weaknesses and develop consensus interpretations regarding regulatory utility. With such attributes, they would be highly valuable for advancing clinical investigations when (1) the human relevance of animal toxicology findings are suspect, and (2) such animal findings are currently poorly monitored in usual clinical practice.

In its Critical Path Document, the FDA has indicated that "there is currently an urgent need for additional public-private collaborative work on applying technologies such as genomics, proteomics, bioinformatics systems, and new imaging technologies to the science of medical product development." As a result on March 16, 2006, the FDA and the Critical Path Institute announced the launch of the Predictive Safety Testing Consortium with such a role in mind (http://www.fda.gov/bbs/topics/news/2006/NEW01337.html) to enable pharmaceutical companies to establish a transparent process for sharing knowledge and resources to evaluate and qualify biomarkers that are appropriate for regulatory applications. These efforts have begun to yield potential new biomarkers with cross-qualification data demonstrating their utility as safety biomarkers for regulatory decision making as reported at the 2007 meeting of the Society of Toxicology *(30)*.

6. Conclusion

When drug development scientists apply new technological approaches to generate toxicogenomic data and exercise their professional judgment to make an internal decision from an exploratory non-GLP study to select one compound over another and to not terminate that particular compound's development, the consequences could be very different than if a regulatory scientist viewing the same data set comes to the alternative conclusion to place the very promising drug's development path on development hold. When the data are unambiguous and conclusive there is little risk, but in early stages of technological applications the risks for alternative and conflicting interpretations will be very high and the consequences potentially costly for delaying the delivery of important new medicines to patients in need.

With objectives of fostering internal company decision making using less-established exploratory biomarkers at specific lead optimization and compound selection stages of a drug's development, there is greater freedom to operate than there is at the GLP toxicology testing stages in studies destined to support the safety of clinical testing of that drug to regulatory authorities. For investigative toxicology applications with objectives of generating testable hypotheses using tissues from animals dosed with compounds whose development has been discontinued, there is greater freedom to operate for sponsors than with compounds under active development destined for regulatory review to support the safe conduct of human clinical trials. Although more and better understood toxicology biomarkers would provide great benefits to both drug development as well as regulatory scientists to enhance a shared understanding of relevant and irrelevant compound safety issues, the rate of their evolution and confidence in their full utility in regulatory toxicology studies has been slow.

The recently introduced regulatory guidance has clarified that reporting expectations and requirements for toxicogenomic data generated from animal toxicology studies are dependent on both the level of understanding of the end-point data, as well as the study purpose. This guidance has opened the door for industry scientists to a high degree of freedom to explore the merits of genomic profiling with samples from early non-GLP internal decision-making compound selection studies. There remain huge opportunities to expand genomic profiling into the highly regulated GLP toxicology studies that serve the primary purpose of supporting clinical investigations. However, lingering data submission concerns around interpretations of ambiguous data that are not yet fully qualified, balanced against fears of fraud and nondisclosure accusations, continue to impede full implementation of toxicogenomic profiling into safety assessment GLP phases of drug development. Evolution of toxicogenomics and other "omic" technologies into regulatory toxicology studies could be accelerated with important modifications to the expectations embodied in the current FDA guidance document that would limit data submission requirements to only qualified biomarkers pertinent to safety and would eliminate the requirement for genomic data reanalyses beyond the conduct of pivotal follow-up toxicology studies of longer duration.

With time and experience, eventually, toxicogenomic data may be linked to other high-output analyses (e.g., proteomics and metabonomics), in combination with standard toxicology data, to form an integrated assessment of a drug's safety profile. These data will likely be more rationally linked to human genomic, proteomic, or metabonomic data to assess relevance of animal findings to clinical situations. Only time will tell if these technologies will evolve to the point where they may reduce any of the standard toxicology studies currently relied upon for regulatory approval and product registration. For positive changes to occur, at the drug development/regulatory interface the data need to be generated, evaluated, linked to outcome, and openly shared to the extent feasible in public-private collaborative consortia dedicated to biomarker qualifications. Evaluation, acceptance, and qualification will occur in discreet steps. Only subsets of the data will be useful at times. An environment that enables and encourages thorough investigation of the technologies and does not penalize the early adopters would be sensible and far-sighted. Unless changes in regulatory expectations regarding the submission of such exploratory data sets are acknowledged and revisited by regulatory authorities, the pace of legitimate benefit to regulatory applications will continue to inch slowly.

References

1. Schena, M., Shalon, D., Davis, R.W., and Brown, P.O. (1995) Quantitative monitoring of gene expression patterns with a complementary DNA microarray. *Science* **270**, 467–470.
2. Ulrich, R. and Friend, S. (2002) Toxicogenomics and drug discovery: will new technologies help us produce better drugs? *Nat. Rev. Drug Discov.* **1**, 84–88.
3. Food and Drug Administration. (16 March 2004) Innovation or stagnation: challenge and opportunity on the critical path to new medical products. Available at http://www.fda.gov/oc/initiatives/criticalpath/whitepaper.pdf.
4. European Medicines Agency. (23 March 2004) Evaluation of medicines for human use. Discussion paper. The European Medicines Agency road map to 2010:

preparing the ground for the future. Available at http://www.emea.europa.eu/pdfs/general/direct/directory/3416303en.pdf. Doc Ref: EMEA/H/34163/03/Rev 2.0.

5. European Medicines Agency. (22 March 2007) Evaluation of medicines for human use. Innovative drug development approaches. Final report from the EMEA/CHMP-Think-Tank Group on Innovative Drug Development. Available at http://www.emea.europa.eu/pdfs/human/itf/12731807en.pdf. Doc. Ref. EMEA/127318/2007.

6. Kola, I. and Landis, J. (2004) Can the pharmaceutical industry reduce attrition rates? *Nat. Rev. Drug Discov.* **3**, 711–715.

7. Chan, V.S.W. and Theilade, M.D. (2005) The use of toxicogenomic data in risk assessment: a regulatory perspective. *Clin. Toxicol.* **43**, 121–126.

8. Code of Federal Regulations. (2 April 2006) 21 CFR 312.23. Title 21. Food and Drugs, Chapter I. Food and Drug Administration Department of Health and Human Resources, Subchapter D Drugs for Human use, Part 312 Investigational New Drug Application, Subpart B Investigational New Drug Application (IND) Sec. 312.23 IND content and format. Available at http://a257.g.akamaitech.net/7/257/2422/26mar20071500/edocket.access.gpo.gov/cfr_2007/aprqtr/pdf/21cfr312.23.pdf.

9. Food and Drug Administration. (1995) Content and format of investigational new drug applications (inds) for phase 1 studies of drugs, including well-characterized, therapeutic, biotechnology-derived products. Available at http://www.fda.gov/cder/guidance/clin2.pdf.

10. Petricoin, E.F. III, Hackett, J.L., Lesko, L.J., Puri, R.K., Gutman, S.I., Chumakov, K., et al. (2002) Medical applications of microarray technologies: a regulatory science perspective. *Nat. Genet.* **32**, 474– 479.

11. Food and Drug Administration. (March 2005) Guidance for industry pharmacogenomic data submissions. Available at http://www.fda.gov/cder/guidance/6400fnl.pdf.

12. European Medicines Agency. (April 2006) Guideline on Pharmacogenetics Briefing Meetings. Available at http://www.emea.europa.eu/pdfs/human/pharmacogenetics/2022704en.pdf.

13. European Medicines Agency. (January 2007) Workplan for the Safety Working Party. Available at http://www.emea.europa.eu/pdfs/human/swp/15242006en.pdf.

14. Food and Drug Administration. (2006) Recommendations for the generation and submission of genomic data. Available at http://www.fda.gov/cder/genomics/conceptpaper_20061107.pdf.

15. Lesko, L.J., Salerno, R.A., Spear, B.B., Anderson, D.C., Anderson, T., Brazell, C., et al. (2003) Pharmacogenetics and pharmacogenomics in drug development and regulatory decision making: report of the First FDA-PhRMA-DruSafe-PWG Workshop, *J. Clin. Pharmacol.* **43**, 342–358.

16. Wagner, J.A., Williams, S.A., and Webster, C.J. (2007) Biomarkers and surrogate end points for fit-for-purpose development and regulatory evaluation of new drugs. *Clin. Pharmacol. Ther.* **81**, 104–107.

17. Ellinger-Ziegelbauer, H., Stuart, B., Wahle, B., Bomann, W., and Ahr, H.-J. (2004) Characteristic expression profiles induced by genotoxic carcinogens in rat liver. *Toxicol. Sci.* **77**, 19–34.

18. Nantermet, P.V., Xu, J., Yu, Y., Hodor, P., Holder, D., Adamski, S., et al. (2004) Identification of genetic pathways activated by the androgen receptor during the induction of proliferation in the ventral prostate gland. *J. Biol. Chem.* **279**, 1310–1322.

19. Kramer, J., Curtiss, S., Kolaja, K., Alden, C., Blomme, E., and Curtiss, W., et al. (2004) Acute molecular markers of rodent hepatic carcinogenesis identified by transcription profiling. *Chem. Res. Toxicol.* **17**, 463–470.

20. Food and Drug Administration. (May 2001) Guidance for reviewers pharmacology/toxicology review format. Available at http://www.fda.gov/cder/guidance/4120fnl.pdf.

21. Code of Federal Regulations. (26 March 2007) 21 CFR 314.50. Title 21 Food and Drugs, Chapter I Food and Drug Administration Department of Health and Human Resources, Subchapter D Drugs for Human use, Part 314 Applications for FDA Approval to Market a New Drug, Subpart B Applications Sec. 314.50 Content and format of an application. Available at http://a257.g.akamaitech.net/7/257/2422/26mar20071500/edocket.access.gpo.gov/cfr_2007/aprqtr/pdf/21cfr314.50.pdf.

22. Wong, J.S. and Gill, S.S. (2002) Gene expression changes induced in mouse liver by di(2-ethylhexyl) phthalate. *Toxicol. Appl. Pharmacol.* **185**, 180–196.

23. Williams, S.A., Slavin, D.E., Wagner, J.A., and Webster, C. (2006) A cost-effectiveness approach to the qualification and acceptance of biomarkers. *Nat. Rev. Drug Discov.* **5**, 897–902.

24. Goodsaid, F. and Frueh, F. (2006) Process map proposal for the validation of genomic biomarkers. *Pharmacogenomics* **7**, 773–782.

25. Bulera, S.J., Eddy, S.M., Ferguson, E., Jatkoe, T.A., Reindel, J.F., Bleavins, M.R., and De La Iglesia, F.A.. (2001) RNA expression in the early characterization of hepatotoxicants in Wistar rats by high-density DNA microarrays. *Hepatology* **33**, 1239–1258.

26. Hamadeh, H.K., Bushel, P.R., Jayadev, S., Martin, K., DiSorbo, O., Sieber, S., et al. (2002) Gene expression analysis reveals chemical-specific profiles. *Toxicol. Sci.* **67**, 219–231.

27. Hamadeh, H.K., Bushel, P.R., Jayadev, S., DiSorbo, O., Bennett, L., Li, L., et al. (2002) Prediction of compound signature using high density gene expression profiling. *Toxicol. Sci.* **67**, 232–240.

28. Thomas, R.S., Rank, D.R., Penn, S.G., Zastrow, G.M., Hayes, K.R., Pande, K., et al. (2001) Identification of toxicologically predictive gene sets using cDNA microarrays. *Mol. Pharmacol.* **60**, 1189–1194.

29. Waring, J.F., Jolly, R.A., Ciurlionis, R., Lum, P.Y., Praestgaard, J.T., Morfitt, D.C., et al. (2001) Clustering of hepatotoxins based on mechanism of toxicity using gene expression profiles. *Toxicol. Appl. Pharmacol.* **175**, 28–42.

30. Sistare, F.D. and Vonderscher, J. (2007) Impact on drug development and regulatory review of the qualification of novel biomarkers of nephrotoxicity. *Toxicologist* **96**, 444.

Index

Printed in the United States of America